MW00737770

Children and Their Urban Environment

'This provocative book takes on *the* question parents, educators and, indeed, anyone concerned about the human future should be contemplating: how can we foster resilience and creativity so today's children can live well through a century of deep change and challenges we can hardly begin to imagine?'
Dianne Dumanoski, author of *The End of the Long Summer*,
co-author of *Our Stolen Future*

'It always seems like the most inviting neighbourhoods are the ones where kids are outside, playing, walking or even running errands. It means the place is safe and welcoming, and that's good for folks of all ages. Here's to a book that brings this priority to city planning!'
Lenore Skenazy, author of the *Free-Range Kids* book and blog

'This book puts children at the centre of things. It demonstrates that if professionals, politicians, urban designers and those responsible for the spaces that children inhabit actually look at things through children's eyes we can visualize and operationalize dramatic improvements in the quality of life for everyone. This is an important achievement. We now have no excuse for delaying the task of making cities child friendly and, in the process, sorting out climate change, peak oil, energy security and resilience. We can also make our cities more joyful, enjoy nature rather than attack it and just have fun.'
Professor John Whitelegg, Stockholm Environment Institute

'Advocates for children and child-friendly communities will be delighted to have this comprehensive book as a guide and reference. Thank you for giving children a voice in this seminal work.'
Catherine O'Brien, Cape Breton University

'More than just a litany of crises and problems facing children in today's urban environments, Freeman and Tranter set out a broad and evidenced-based agenda for social and cultural change. International in its scope, this book would be a marvelous addition to a wide variety of undergraduate and graduate courses across the social sciences – a must-read for professional urban planners and policy-makers.'
Mark Lowes, Associate Professor, Department of Communication,
University of Ottawa

'As both a father and a researcher, I know how important the urban environment is to children as they progress into adulthood. This book, which is both readable and of high academic quality, will help readers understand the importance of good urban environments in developing our future generations.'
Associate Professor Simon Kingham, Head of Geography,
University of Canterbury, New Zealand

Children and Their Urban Environment

Changing Worlds

Claire Freeman and Paul Tranter

London • Washington, DC

First published in 2011 by Earthscan

Earthscan Ltd, Dunstan House, 14a St Cross Street, London EC1N 8XA, UK
Earthscan LLC, 1616 P Street, NW, Washington, DC 20036, USA

Earthscan publishes in association with the International Institute for Environment and Development

For more information on Earthscan publications, see www.earthscan.co.uk
or write to earthinfo@earthscan.co.uk

ISBN: 978-1-84407-853-0

Typeset by Domex e-Data Pvt. Ltd., India
Cover design by Rob Watts

A catalogue record for this book is available from the British Library

Library of Congress Cataloging-in-Publication Data

Freeman, Claire.
Children and their urban environment : changing worlds / Claire Freeman and Paul Tranter.
p. cm.
Includes bibliographical references and index.
ISBN 978-1-84407-853-0 (hardback)
1. Cities and towns–United States. 2. Urbanization–United States. 3. Urban economics–United States. 4. Sociology, Urban–United States. 5. City and town life–United States. 6. Urban policy–United States. I. Tranter, Paul J. II. Title.

HT123.F75 2010
307.1'16083–dc22

2010037835

At Earthscan we strive to minimize our environmental impacts and carbon footprint through reducing waste, recycling and offsetting our CO_2 emissions, including those created through publication of this book. For more details of our environmental policy, see www.earthscan.co.uk.

Printed and bound in the UK by MPG Books,
an ISO 14001 accredited company.
The paper used is FSC certified.

This book is dedicated to our children
Nick, Kathy, Shaun, Steven and Colin.
They have helped us see cities through the eyes of a child.

Contents

List of figures, tables and boxes *ix*
Foreword *xiii*
Acknowledgements *xv*
List of acronyms and abbreviations *xvii*

Part I—Changing Environments, Changing Lives

1 Children's Life Worlds: Adapting to Physical and Social Change 3
2 Same World – Different World 19

Part II—Activity Spaces

3 Home 39
4 School 57
5 Neighbourhood 77
6 City Centre 93
7 Service Space 115
8 Cultural Spacc 137
9 Natural Space 159

Part III—Making a Difference: Creating Positive Environments for Children

10 Accessing Space: Mobility 181
11 Design 203
12 Professionals and Children: Working Together 223
13 Conclusion: Children's Play and Resilient Cities 243

Index *257*

List of Figures, Tables and Boxes

Figures

1.1	Children with disabilities can experience space differently	4
1.2	The redevelopment of Darling Harbour, Sydney	5
1.3	Children take advantage of a patch of grass to play in the city centre	11
1.4	In cities such as Amsterdam the child's right to the city is prioritized	12
2.1	Learning to negotiate traffic is a risky endeavour, but a skill that children have to learn	26
2.2	Children are growing up in an increasingly risk-averse society	27
2.3	Children are keen to interact with their environment and with each other	32
2.4	Playing hide and seek in the trees	33
2.5	Child playing on a pole on the Washington Metro	34
3.1	These homes were not designed for children, but the reality is that children live in them	41
3.2	Well-planned homes with good open relationships with the street encourage social contact and play between children and young people	42
3.3	When Japanese children start primary school at the age of six they receive two presents	50
3.4	Unfortunately, many new homes are being built that are alienating in their design	53
4.1	An ideal school ground from a child's perspective?	59
4.2	Risk compensation	62
4.3	Planting rice in a small rice field set up at the corner of the school ground	68
4.4	Play materials introduced into a Sydney, Australia, school ground to encourage social creative play	70
4.5	Drawing of Orana School ground, Canberra, Australia, as a child would like it to be	72
5.1	Children play anywhere and everywhere, and a good neighbourhood encourages this	78
5.2	Celebrations and festivals are an important part of public life and bring communities together	79
5.3	Homes like this one in Dunedin, New Zealand, contribute to social isolation	81
5.4	Good neighbourhoods that integrate built and natural features are aesthetically appealing	86
5.5	'If I was a town planner'	90
6.1	Buskers in public spaces make an important contribution to the cultural life of the city	98
6.2	The Buckets are a well-loved feature in the cityscape of Wellington, New Zealand's capital city	99

6.3 Colonizing leftover space, under the Arches, South Bank, London 101
6.4 Children gravitate towards public art that encourages interaction 102
6.5 The Bull in the Bull Ring Shopping Centre is an imposing and historically relevant piece of public art 103
6.6 The Lions in Trafalgar Square, London, have acted as a climbing magnet for generations of children 104
6.7 The Steam Sculptures beg to be climbed 105
6.8 Children's designs incorporated within a high-quality artistic, yet functional, tree protector 106
6.9 Drawing by a fifth year Canberra school child of the city centre as a good place to be 112
7.1 A Sydney school with a high fence, topped with barbed wire 116
7.2 Adults as well as children can enjoy conventional playgrounds 118
7.3 A Liberty Swing for use by children in wheelchairs, separated from other park users by a high fence 121
7.4 Farmers' markets provide excellent places for children to connect with a very active public space 125
7.5 Children being planes in the Science Museum in London, UK 128
7.6 Drawing of things that would make a museum a good place for children 132
8.1 National pride is an important part of spaces in Washington, DC 138
8.2 Children interact with the city in diverse ways 139
8.3 Schools can play an important role in supporting children's cultural identities 144
8.4 These children attending a mosque, although all Muslim, represent a range of ethnic, language and cultural groups 148
8.5 Children in traditional dress taking part in the St Patrick's Day Parade 151
8.6 Birthdays are an important celebration in many children's lives 152
9.1 Having fun in the mud 160
9.2 Catching tadpoles using adapted milk containers 162
9.3 Gardening provides contact with nature, enjoyed across all age groups 164
9.4 Even formal parks provide playful contact with nature 167
9.5 Children value the everyday contact with nature that common, usually overlooked garden species such as this stick insect provide 171
9.6 Child using the interactive outdoor displays at the London Wetland Centre 173
10.1 Children using the street as a play space 182
10.2 A walking school bus to P. L. Robertson Public School, Ontario (Canada) 186
10.3 Proportion of British 10- and 11-year-olds able to undertake activities unaccompanied 188
10.4 Mode of travel to school for children aged five to nine in Australia 188
10.5 A school bus collecting children from John W. Ross Elementary School, Washington 192
10.6 Designating Home Zone areas is one approach to reclaiming streets for children 195
10.7 A suburban street in Melbourne 195
10.8 Active transport is also important for children 197

11.1 This development prioritizes pedestrian child-friendly spaces in its design 211
11.2 Good neighbourhood design encourages active multi-age group play 213
11.3 Although not an especially appealing housing design, the courtyard is nonetheless well used 214
11.4 Retrofitting urban areas to prioritize pedestrian rather than car accessibility 220
12.1 An example of the box city approach 224
12.2 Methods used need to be child appropriate 230
12.3 For many years, working on joint projects with local school children formed an integral part of the education of planning students 232
12.4 Maps can be used as a way of understanding children's life worlds 236
12.5 Drawings can give good insights into children's views on places 237
13.1 The energy crisis is one that will be a significant part of children's future 245
13.2 Playful children mixing with adults in a reclaimed Times Square, New York 248
13.3 Breaking down barriers between children and those responsible for 'managing' urban environments 250
13.4 Alternative uses for cars in the city 252
13.5 We can build on children's desire to take a risk with play 253

Tables

1.1 Key themes in the construction of childhood 9
1.2 Social and space-based paradoxes around children 13
1.3 Some of the diversity of family types, homes, residential areas and functional places that children experience in the city 14
2.1 Changes in cities and lifestyles affecting children 22
3.1 Children's housing needs 43
3.2 Children's views on what makes a great home 53
4.1 Children's views on what makes a great school ground 71
4.2 Children's views on what makes a great classroom 72
5.1 Alberta Premier's Council, Canada: Community self-assessment checklist 80
5.2 Phases in European house development 85
5.3 New housing development types 87
5.4 Children's views on what makes a good neighbourhood 90
6.1 City-centre facilities 94
6.2 Features and facilities close to Coin Street 110
6.3 Children's views on what makes a good city centre 111
7.1 Children's views on what makes a good museum 131
8.1 Population composition for three schools in South Auckland, New Zealand 144
8.2 Culturally significant events 153
9.1 Urban green spaces 172
9.2 Types of green spaces easily accessible to children living at a random sample of addresses 174

11.1 Child-friendly urban design principles 206
11.2 Urban design principles used by organizations and cities 207
11.3 Relationships to be considered in housing design 215
11.4 Design principles for children's outdoor space 219
12.1 The seven realms of children's participation in city design and planning 228
12.2 Built environment and related professions whose activities impact upon children's lives 229
12.3 The benefits of participation 234
12.4 Access modes commonly used to facilitate children's participation 238

Boxes

2.1 Affordances in Washington, DC 34
4.1 Two school grounds in Seattle, Washington 59
4.2 Washington Elementary Schoolyard: Stimulating environmental learning 66
5.1 SickKids Toronto checklist: Location, location, location 88
7.1 Boundless Playgrounds 120
7.2 A more child-friendly hospital 127
7.3 'Kids in Museums Manifesto' 129
7.4 Prisons as a children's service space 130
8.1 Children's experiences in a mosque 148
10.1 How walking to school can save parents time 184
10.2 Three days to a more child-friendly city 194
11.1 *Children's Manifesto*, Bologna, 1994 205

Foreword

With the advent of the United Nations Convention on the Rights of the Child (UNCROC) and its systematic monitoring, the rhetoric of children's rights has become universally framed. Never before has there been a time when children have been under such a spotlight. Yet, despite this upsurge of interest in the lives and well-being of children, many children still find themselves at odds with their societies, denigrated to being little more than citizens in waiting. The values that at one time suggested that children should be seen but not heard have yet to be fully dissipated. Too much tokenism still exists, such that where participation has been advocated and applied, outcomes become measured not on the results that change those circumstances that bind children, but in the process of taking part itself. It would seem that children's active engagement is often no more than a tick-box exercise, which once recorded enables decisions to be taken regardless of what has been expressed. Listening becomes the end, not the start, of a journey of enlightenment. Another scenario is where well-meaning adults define themselves as the authentic witness to children's voices, but in doing so take charge of those situations in which children could readily and willingly get involved.

Within this book we are provided with a refreshing insight into the multifarious lives of children and their divergent childhoods, set in parts of the developed world. Its starting point is that societies are not static and that children's position within them is constantly evolving, partly as a result of their own volition, but mostly through actions and contexts beyond their control. The immediate environmental challenges that now face humankind have particular resonance for those whose lives are just beginning. Yet, major social and political barriers often impinge upon how children and their competences are regarded, suppressing their ability to take part in decisions that not only affect their 'here and now', but also their futures. As a result, a common misconception is that children cannot take part in projects or in their communities in ways other than being child-like. However, this need not be the case. *Children and Their Urban Environment* presents compelling evidence on how, if expectations are opened up, the interests of children need neither collide nor be subservient to those of their fellow citizens, and how children and adults, by working together, can bring about stronger, more sustainable and healthier communities.

By examining children's worlds through their own eyes, new perspectives are provided on how children see and experience the world around them. These visions emphasize that children are creators, not just consumers of places, and that in the absence of strategies that empower children, communities will lack those essential moral and democratic ingredients that enable children to truly flourish. The authors demonstrate that space at various scales – whether home, school, neighbourhood, city centre or cultural space – need not be associated with practices of exclusion and frustration, but (through positive engagement) can provide opportunities that include, support and nurture the well-being of children. Taken as a whole, this book provides a powerful polemic on the benefits of regarding children's active participation as an entitlement towards energetic citizenship, and not a privilege bestowed by enlightened adults.

Wrapped up in such debate are notions of power and powerlessness. Here, this book is unequivocal and its advice is straightforward and most welcome. Enabling young people to fulfil their potential as equal participants will involve some adults relinquishing what they regard as rightfully theirs. Deep-rooted hegemony of this kind has hampered the process of inclusion over decades and centuries. It is something that cannot be broken down through the recording of piecemeal interventions, no matter how successful these may be. What is required is the need to change social and political values, such that young people's needs become grounded through their own contributions within structures that value and welcome their participation without fear of loss or *emasculation*. This will come about through the active dissemination and application of good practice, based upon examples that highlight the enrichment that is achieved within these new landscapes of opportunity.

Presented in three parts, with a broad international focus, Part I explores the changing nature of childhood and how emerging environmental challenges are beginning to impact upon children in a variety of ways. Here, childhood is not depicted as a unitary norm, but as a diverse amalgam whose various conditions lead to a multiplicity of issues. Consensus is evident, however, around the many ways in which children are disenfranchised within their localities. In Part II the authors use the views of children to help them articulate how everyday spaces and places afford or negate opportunities for play, socialization, learning and living. With these lessons in mind, Part III sets out a broad agenda for change that, if adopted, would strengthen the active social commitment of children within their communities, leading to them becoming positive agents in the making of their own sustainable futures. Partnership working of this kind will not be easy. It will involve the dismantling of those structures that privilege the status and eminence of adults. However, if such an alliance can be achieved, then environments will be more resilient and better spaces for living for all.

It is with great pleasure that I am able to recommend this book to you.

Hugh Matthews
Professor of Children's Geographies
University of Northampton, UK
8 September 2010

Acknowledgements

We are immensely grateful to a number of individuals and organizations whose support, advice and encouragement made this book possible. We would like to thank them all for their time and involvement, which was so willingly given.

Our thanks go to:

- the reviewers, whose advice on our proposal helped to refine our book into a much more inclusive one;
- Margaret Finney, for a most necessary proof-read and for her advice on the written draft;
- Julie Kesby, for her continued encouragement and valuable editorial advice;
- all of those who kindly provided the photographs used in this book and to the children who feature in them;
- the children and teachers from Year 5/6 at Forbury Primary School Dunedin, from Year 4/5 at Blue Gum Primary School and Year 5/6 at Ngunnawal Primary School in Canberra, who enthusiastically shared with us their views on the city and their drawings; also the children and parents from Al-Huda Mosque Dunedin;
- our colleagues and our universities, the University of New South Wales at the Australian Defence Force Academy (UNSW@ADFA), the University of Otago and the University of Queensland, who supported our endeavours;
- the two ethics committees at the University of New South Wales and the University of Otago, who provided approval for our research with primary school children;
- our publishers, who have supported and advised us throughout the writing process;
- Sue and Nick, who provided moral support and encouragement, helped with the choice of photographs, and provided a sounding board for topics covered in the book.

List of Acronyms and Abbreviations

ADHD	attention-deficit/hyperactivity disorder
ADM	adult-dependent mobility
BBC	British Broadcasting Corporation
BMI	body mass index
BSL	British sign language
CBD	central business district
CCTV	closed circuit television
CEO	chief executive officer
CIM	children's independent mobility
CO_2	carbon dioxide
ESRC	Economic and Social Research Council
GDP	gross domestic product
GIS	geographic information system
GPS	global positioning system
IUCN	World Conservation Union (*formerly* International Union for the Conservation of Nature)
LEED	Leadership in Energy and Environmental Design
NGO	non-governmental organization
OECD	Organisation for Economic Co-operation and Development
PIA	Planning Institute of Australia
SSSI	Site of Special Scientific Interest
SUV	sports utility vehicle
UK	United Kingdom
UN	United Nations
UNCED	United Nations Conference on Environment and Development
UNCROC	United Nations Convention on the Rights of the Child
UNESCO	United Nations Educational, Scientific and Cultural Organization
UNICEF	United Nations Children's Fund
UNSW@ADFA	University of New South Wales at the Australian Defence Force Academy
US	United States
4WD	four-wheel drive
WSB	walking school bus
WWF	World Wide Fund for Nature (*formerly* World Wildlife Fund)

Part I

Changing Environments, Changing Lives

1

Children's Life Worlds: Adapting to Physical and Social Change

> Since childhood is one of the few absolutely universal experiences, it is not surprising that people have an inward picture, even though it may never be articulated, of an ideal childhood. We may use it to reshape our own memories, we may try to recreate it for our own children. (Ward, 1978, p2)

Childhood

Childhood is generally accepted as being primarily a social construction. Its character is moulded by the norms of the society into which the child is born. These norms dictate matters as diverse as the influence that gender will have on the child's 'value' and life opportunities, the likelihood of the child being part of a nuclear or extended family, the child's spiritual dimensions, and even matters as pragmatic as what the child will eat for breakfast. Children are born into a society which will nurture them, sustain them and protect them, as well as neglect them, frustrate them, constrain them and provide them with their social and individual identity. Children possess their own character; but the society within which they are born and raised will impact upon how this character will develop and the extent to which children will be supported or obstructed in their desire and ability to become themselves. Adapting to and finding their way in society is one of the greatest of all accomplishments a child will make.

Adapting to society is one of the greatest accomplishments a child will make

How do children become socialized into society? Issues around the social development of children have intrigued those seeking to advance a philosophy of childhood. Early writers on this topic include luminaries such as Piaget (1954), Durkheim (1979), Erikson (1993) and Bowlby (1999, originally published in 1969), to more recent and, in some cases, lesser-known childhood thinkers such as Liedloff (1977), Postman (1994), Frones (1995), Spock and Needlman (2004), Stanley et al (2005), Donahoo (2007), Furedi (2008) and Leach (2009). What this disparate group has in common is the desire to understand and frame childhood. In the case of Piaget: to understand the conceptual and physical development of the child; in the case of Leach and Spock: to shape parental behaviours in a way that creates an 'ideal' family setting in which the child can develop.

While society is clearly important to children, it is only half the equation. As well as being shaped by their social world, children are shaped by their physical world: the places and spaces in which they grow up. This spatial world can be highly influential, as indicated in the following example where children may experience a similar family structure but live in spaces that impact very differently upon their life experiences. Two children are growing up in a two-parent, three-sibling family where the breadwinner works as a bus driver. One child lives with their family in a deprived, vandalized 'sink estate' in central London, the other on an established council (government housing) estate with low residential mobility in leafy Epsom, on London's outer fringes. To take a second example, the experiences of a child growing up in an apartment above a shop in one of Toronto's multicultural inner suburbs such as Agincourt with its large Chinese community would be quite different from a child growing up in an upmarket condominium in Toronto's desirable South Beach suburb. These suburbs can be physically quite close; but the lifestyles and influences associated with them can be quite different. The same space can also be experienced very differently by children of differing abilities. An environment that may be welcoming and accessible to a physically able child may be experienced as challenging and isolating to a child with a disability (see Figure 1.1). Space matters: place, like society, shapes and influences behaviours, the spirit, sociability, opportunities, play, health, independence, physical and mental well-being, and even happiness. Yet, space has been given little attention in childhood literature; it has been taken as a given, allocated background status or reduced to relatively simplistic terms, such as healthy or unhealthy homes, environments which are good or bad for play, natural environments or

Figure 1.1 Children with disabilities can experience space differently

Source: Gretchen Good

sterile environments. Space acts as the nexus where society and place converge in the child's life.

Socially Determined Space

Space is complex: it varies not only at the larger scale between different cities, or parts of the city, but at the micro-scale, as in different parts of the street or home. These variations impact upon children in the most immediate ways. The design and culture of the street determines whether the child can cross a street to meet friends on the other side, while the height and design of the front garden fence will influence whether the child can see and interact with the 'outside' world or not. Space is also dynamic: its physical properties change. A quiet neighbourhood street can become a 'rat run' for commuting traffic, precluding street play and accessing neighbours on the other side. A neighbourhood that was once full of large upper middle-class homes can become a place of flux as family homes are subdivided into rentable rooms that house predominantly mobile populations. Not just buildings but the spaces around them change: a Victorian park can become a vandalized 'no-go area' characterized by minimal use, but with the right inputs can be revived into a safe community meeting and play place (see Figure 1.2).

Tied up with physical change is social change. Where once it was seen as safe for children to be out and about at the park, use public transport and walk to school, some societies now see these same activities and journeys as dangerous and needing adult accompaniment. This was evident in the outcry directed at Lenore Skenazy (2009), author of *Free-Range Kids*. She wrote a

Figure 1.2 The redevelopment of Darling Harbour, Sydney, has created exciting, varied public spaces enjoyed by children and their families

Source: Claire Freeman

piece for the *New York Sun* in which she described how her nine-year-old son used the New York subway alone. The article engendered some vehement responses in which many readers castigated her for her irresponsibility, with some going so far as to class her as 'America's worst mum'. Yet, if we look at the influential book by Colin Ward, *Child in the City* (Ward, 1978), published some 30 years ago there are several photographs showing exactly that: children using public transport without adults, some of the children being considerably younger than nine. Indeed, many children still take public transport alone to attend school. If you stand in the central business district (CBD) in Sydney, Australia, any school-day morning, multitudes of children from around age 11 can be seen emerging from buses and trains crossing the city to their usually private or selective schools – a seemingly acceptable undertaking. If Lenore Skenazy's son had been dressed in a school uniform heading to a New York City centre school, would there have been the same outcry? Would groups of 11-year-olds taking the train to spend the day hanging about in Sydney's CBD be as accepted as their school-uniform-wearing mid-week counterparts? We'd guess not: same journey, same geographic destination, but different societal associations. Furthermore, would those same school-going 11-year-olds who use public transport in a large metropolitan centre with assurance be free to wander around their own suburban streets unfettered? Evidence also suggests not. So what is going on? Why is space so imbued with social rules and behavioural codes, and what is it that determines these? Is it something intrinsic to space itself; is it something intrinsic to the society and to the societal meanings placed on space; or is it some combination of these? These are some of the questions explored in this book.

In their planning and design, most spaces ignore children as users

Space in all its variations – home, school, street, bus stop, shop, soccer pitch, playground, health clinic, library, garden, city centre, public square, to name but a few – forms an integral component of the child's world. Yet in their planning and design most of these spaces ignore children as users. This was brought home very clearly in research that Freeman and Aitken-Rose (2005a, 2005b) undertook with local authorities in New Zealand. In this study planners were asked about how and where they took children into account in planning. The responses invariably were that planners considered children in planning for recreation, mainly playgrounds and sports, and in planning for education, schools and crèches. Planners were particularly pleased at being able to recount how they were taking young people into account in the processes around planning for skate parks. For what other sector of society would one 'minority' and highly gendered sporting activity be seen as meeting general social and sporting needs? Also, what other sector of society would have their needs ignored when planning homes, streets, roads, shops, health and leisure facilities, transport and infrastructure? Planners also held simplistic notions of children, characterized by a concept of a universal child, homogeneous and undifferentiated: a boy of a certain age is assumed to be almost definitely interested in skateboarding. Children are not like that; their lives are not like that. We do not plan for a 'universal, uniform prototype adult'. Like adults, children reflect the infinite variety of life, culture, age, race, gender, experience, character, level of ability, likes and dislikes, and are differentially affected by the environments and processes of environmental change. In the city children do more than recreate and be educated (as the planners believed). They use the whole city. As Colin Ward said: 'I want a city where children live in the same world as I do ... [where if] the claim of children to share the city is admitted, the whole environment has to be designed and shaped with their needs in mind' (Ward, 1978, p204).

The whole environment has to be designed with children's needs in mind

The New Zealand planning research acted as an impetus for this book (Freeman and Aitken-Rose, 2005a, 2005b). It revealed the need to consider children in all societal variations, across different types of environments and with recognition of difference in children's experiences and use of space. Seeking to understand this relationship is, however, only a recent occurrence in the annals of social and environmental history.

Changing Contexts and Conceptualizations of Children's Lives

There has been considerable and growing interest in children over the last 30 or so years. A number of texts have explored the changing position of children in society. It is a task made difficult by the paucity of recognition that children are given in historical records. In part, this is symptomatic of the fact that for much of history children were not necessarily seen as separate or distinguishable from adults. Some authors focus on the negative permutations of childhood to explain its relative absence. Lloyd deMause (1974) goes so far as to state that 'the history of childhood is a nightmare from which we have only recently begun to awaken' (in Jenks, 2005, p326), and then proceeds to paint a most depressing picture of childhood historically as a period of abuse and misuse. In his fascinating book *Childhood in World History*, Stearns (2006) again refers to the problem of writing on history given the lack of records, pointing to the futility of trying to generalize childhood with its multiple manifestations across time and in different societies. In the midst of this uncertainty, what is undisputed is that, historically, childhood was a time of great uncertainty: death of children was common, families were large and the investment of time and resources by families in any individual child was minimized, at least until the child had proved some likelihood of survival into adulthood. Authors such as Stearns demonstrate that interest in childhood as being worthy of dedicated study is a relatively recent phenomenon, one only coming to the fore in the latter half of the 20th century.

The child's place and role in society has been the focus of robust debates. Positions within this debate can be held strongly, with children being seen as everything from vulnerable innocents, blank canvases, young adventurers and deviants to competent social actors. These differing positions have informed markedly different approaches to the exploration of childhood and children's environments. Postman (1994) writes of the 'disappearance of childhood', lamenting the loss of innocence as children are too soon inveigled into the not-so-innocent worlds of adulthood. More recently, Donahoo (2007) bemoans what he calls the 'idolising of children' where the pendulum has swung far in the direction of exhorting the creation of impossibly 'ideal' childhoods, an ideal that neither children nor parents can comfortably or realistically achieve. In a similar vein, Furedi (2008) decries what he sees as the rise of 'paranoid parenting': the seeing of danger in all elements of childhood such that any sense of perspective on real dangers is lost. The outcome of 'paranoid parenting' is the continual adoption of risk-averse behaviour: behaviour that is counter-intuitive as children fail to develop the skills necessary for survival or societal competence. Risk-averse behaviour is also the focus of the work of Gill (2007), who attests that the concern about health and safety is out of all proportion to the real levels of risk that children are likely to encounter. As a counterpoint, there is a developing literature that sees children as competent social actors who not only react to social and environmental circumstances, but also use their own agency and autonomy to shape them.

A developing literature sees children as competent social actors

Where does this leave those of us concerned with the well-being of children and childhood when such often vehement and polarized debates emerge on the realities of childhood, and what does it have to do with a book on space and place? We have a plethora of often conflicting exhortations on children's well-being with regard to their societal and place-based relationships. Within this profusion lie some universals, such as the need to be safe, to socialize, to develop personally, to live in suitable housing, to be healthy, to play and to belong to a nurturing family. However, the way in which these 'universals' are manifested and the form they take will be socially and environmentally determined according to what is acceptable and appropriate in different places and times. These invariably will and should be, to some extent, culturally and place specific. For example, the design of homes in suburban US and high-density inner-city Paris will be different; but within these homes the right of children to experience dry, warm, ventilated conditions as opposed to damp respiratory disease-inducing conditions is equally applicable. The right to have safe areas for outdoor play again will take a different form in small town Sweden, where the child may freely traverse a wide area on the rural fringe (Kytta, 2004), compared to the much more confined play courtyards for children living in high-density housing in Hong Kong.

The appropriateness of space and its activities are determined by reference to how childhood is conceptualized. If we look at key themes that have emerged in literature on child development, three stand out (see Table 1.1). All will influence the ways in which children lead their lives and the values given to the activities they pursue. For example, play under theme 1 ('Childhood as protected innocence') would be valued as spontaneous child-directed play and could take the form of groups of children on their own building dens in the garden or local park from natural materials such as logs and grass. Under theme 2 ('Childhood as preparation for adulthood'), play is valued as a learning experience helping to prepare children for their adult lives. Play becomes more an opportunity to develop life skills. Building huts takes place in the form of a school grounds-based exercise where parents are involved in supplying any necessary building materials and guiding children in their use of tools to construct the huts. Finally, in theme 3 ('Childhood as equal value to adulthood'), children's play is valued in its own right and may even be seen as equivalent to a similar adult activity. A 'planning' favourite in this regard is getting children to build a 'box city' in a council-designated space. While the activity is inclusive of play elements, its primary function is to feed children's views into wider adult planning processes. In these three examples, the activity – creating a built structure – is the same, but the place and the level of external input changes as

Table 1.1 Key themes in the construction of childhood

Key themes	Manifestations
Childhood as protected innocence	Protection and separation
	Innocence
	Lack of knowledge
	Free play
	Freedom
	Protection and nurturing rights: food, shelter, love, security
Childhood as preparation for adulthood	Education
	Social engagement
	Civic values
	Age-related privileges as stepping stones
Childhood as equal value to adulthood	Adoption and application of appropriate age-relevant adult values
	Exposure to, and engagement with, media and technology
	Rights: voting, participation, personal space

the activity becomes more formalized socially and in space. The process of adult colonization of children's activities is a major theme in the lives of many urban children today and will be revisited in the book. Analogous to the rising interest in children's societal relationships, albeit of a later appearance, has been the rising interest in children's environments.

Children's Use of Space

The growing interest in 'children and society' as a focus for study has been paralleled by developing interest in children's environmental relationships. This interest has been particularly associated with the disciplines of geography, planning, architecture and housing. Within this multidisciplinary setting, with its emerging literature, there have been two discernable trends. In the first, the city is seen as a backdrop against which children's lives are played out. From the early 1990s onwards, a number of major texts were released focusing on children and the city – such as *Children of the Cities* (Boyden with Holden, 1991) and *Cities for Children: Children's Rights, Poverty and Urban Management* (Bartlett et al, 1999) – and includes books arising out of the United Nations Educational, Scientific and Cultural Organization's (UNESCO's) Growing Up in Cities programme (UNESCO–MOST, undated), such as *Growing Up in an Urbanizing World* (Chawla, 2002) and *Creating Better Cities with Children and Youth* (Driskell, 2002).

These books draw attention to cities as important determinants of children's lives and as places where many children's well-being has been severely comprised. A particular concern has been with the lives of children in rapidly urbanizing countries. These books, however, also reveal the vitality of children's lives, where to merely focus on their deprived socio-economic circumstances is to omit recognition of the vibrant, creative ways that children function in society and construct meanings of place regardless of their physical conditions.

The second trend has been in developing better understandings of how children use space, and how space itself changes both socially and physically. This interest sees its genesis in the work of a small cohort of UK- and US-based researchers. Three key early books were *Child in the City* (Ward, 1978), *Children's Experience of Place: A Developmental Study* (Hart, 1979) and *Childhood's Domain: Play and Place in Child Development* (Moore, 1986). Ward's book challenged the accepted adult centric-ness of urban life. He explored the different relationships that children have with urban space, showed how places often overlooked in the adult world (the micro-spaces of the city, such as footpaths, walls and kerbsides) can be important places of refuge, socializing and play for children. The photos in his book vividly evoke children's deep and often playful use of space. The works of Hart and Moore broke new ground with their focus on children's use of space. In this they diverged from conventional writing on children's environments by taking their lead from the children themselves. It was the children who directed much of the subject matter for the books, and who allowed the authors into their own spatial worlds, their meanings and realities. Now such two-way researcher/child relationships are more normal and encouraged. However, even Hart and Moore would have difficulty getting the ethics approvals now required by researchers, particularly to spend their time following children around as they play in order to study them. While the ground had been broken for children's spatial studies by the 1980s, it would be some time before children's geographies and place-based studies really took off and focused attention began to be given to children's lives and to the development of the now mushrooming 'children's environments literature'.

Emerging Concerns

The interest in children and space has been driven by a number of focused concerns. An early focus was the relationship between children and nature prompted by concerns that children, especially city children, were becoming alienated from the natural world. Reduced access to the natural world was seen to be linked to the alienation of the natural world from city life and to children's own declining outdoor access (see Figure 1.3). In particular, outdoor exploratory play was seen as declining and being replaced by a process referred to as the 'domestication' of play. Play and the need for the provision of challenging, active play opportunities became a key area of disquiet on the part of many childhood advocates and professionals. Attention began to be given to children's play, pushed by two, in some ways deeply oppositional, drivers. The one driver is that children's play should be a priority, but is rooted in a belief that play should be provided for. Playgrounds, formal activity spaces and equipment feature heavily in this regard. It is a movement that has, to some extent, been seized by the safety and commercial movement as play has become commodified, structured and sanitized. The second driver is the need to recapture free exploratory play in natural surrounds.

The public domain has become an area of contestation for children

Parallel to the concerns around play and independent mobility are those around children's rights to the public domain. The public domain has become an area of contestation for children as their relationship to public space is questioned. Children have become increasingly relegated to 'child spaces' in the

Figure 1.3 Children take advantage of a patch of grass to play in the city centre, but their play, while outdoors and seemingly spontaneous, forms part of a supervised holiday programme outing

Source: Claire Freeman

city (playgrounds, skate parks, school grounds) and seen as increasingly unwelcome in parts of the city – for example, newly commercializing spaces (shopping malls) as well as places traditionally used for play (woodlands, parklands, around rivers, and wild and undeveloped spaces). Their presence on the street, in public spaces and in natural spaces (traditionally major social and activity sites for children) has become a source of disquiet; indeed, children's visibility in many urban areas is conspicuous by its absence. Another area of decline noted across many countries in the developing world, notably the UK, the US, Australia and New Zealand, is that of independent mobility. Fortunately, this decline is not universal and is less substantial in some European countries, especially the north European Scandinavian countries. Where the decline has occurred, its ramifications have been profound and associated with major changes in children's social interactions, especially at family level as families engage in ever-increasing levels of child-accompanied activity. Children's independent mobility has been replaced by adult-dependent mobility.

The decline in mobility has caught the attention of the medical profession, being seen as a contributory factor to the rising problems of obesity, diabetes and other diseases associated with more sedentary lifestyles. The causes of this decline are complex, but two elements stand out: first, the removal from the street due to concerns about traffic; and, second, concerns about children's safety with regard to other members of society. The retreat from the street has been the focus of a number of studies (Valentine, 2004), as well as the Economic and Social Research Council (ESRC)-funded project on 'The street as third space' (Matthews et al, 2000). Such has been the potency of emerging safety concerns that a new genre of study has emerged: the 'geographies of fear' – a genre in which children feature as a prominent focus both as vulnerable, needing protection in increasingly unsafe environments, and, conversely, where the presence of children and young people (especially groups of teenagers) contributes to some members of the public feeling 'unsafe'.

These differing and diverse concerns – access to nature, play, freedom and independence – have been brought together in the child-friendly cities literature. This literature adopts a broad approach to children's urban relations inclusive of health, transport, play, design, participation, social engagement, environmental interactions and the move towards globalization. The child-friendly cities research seeks not only to understand spatial relationships, but to use this knowledge to redress inequities, restore children's rights and work towards the enhanced well-being of children in the urban realm (see Figure 1.4). The child-friendly cities movement is a global movement. Its many dimensions include the United Nations Children's Fund's (UNICEF's) Child Friendly Cities project, regional initiatives such as the Asia Pacific Child Friendly Cities Network, and individual child-friendly strategies and commitments at local government level. The movement reflects and builds on the momentum of other global environmental directions, notably sustainable cities, green cities, slow cities, Agenda 21 and other programmes – projects and approaches that prioritize good urban design, environmental coherence and social equity.

Towards a Better Understanding

The stage seems to be set for an enhanced urban form for children. There is better understanding of, and commitment to, building urban relationships,

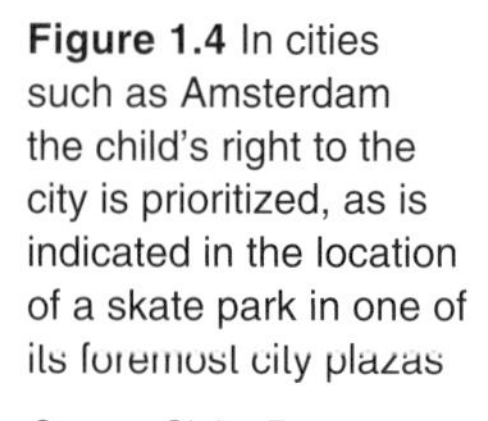

Figure 1.4 In cities such as Amsterdam the child's right to the city is prioritized, as is indicated in the location of a skate park in one of its foremost city plazas

Source: Claire Freeman

promoting children's rights, developing partnerships, and recognizing difference in culture, ethnicity, gender and disability. Yet, challenges persist and continue to overwhelm. Those seeking to enhance children's urban spaces are too often frustrated in their attempts at environmental improvement and paradoxes remain. Quortrup (1995) identified a number of social paradoxes which demonstrate that progress in some areas is counteracted by divergent realities, where not only is progress limited, but in some cases conditions for children are worsening. These paradoxes not only occur socially, but also spatially. Some of these paradoxes are indicated in Table 1.2 and are themes addressed in the book – for example, the paradoxical role of cars in urban society, the unknown role of housing and the contestations around legitimate play space.

To move forward requires understanding the paradoxical, contradictory and dynamic positions of children in society. Social differences and social changes are also reflected in the spatial differences in children's lives. This difference affects both those seeking to understand children and to accommodate

Table 1.2 Social and space-based paradoxes around children

Social paradoxes
Adults want and like children, but are having smaller families, while society is providing less space and time for them.
Adults argue that it is good for children and parents to be together, but more and more they live their everyday life apart.
Adults appreciate the spontaneity of children; but children's lives are more and more organized.
Adults state that children should be given first priority; but most economic and political decisions are made without consideration for children.
Adults agree that children should be given the best start in life; but children nevertheless grow up in less affluent parts of society.
Schools are generally seen by adults as important to society; but children's contribution to knowledge is not recognized as valuable.
Childhood is valuable to society in material terms; nevertheless, society leaves the bulk of expenses to parents and children.
Space-based paradoxes
The importance of the home is stressed in children's well-being; yet in society, children are the ones most likely to live in overcrowded homes.
The home is the most important space in children's lives, yet little is known about how children use their homes, how space is allocated and what children's spatial needs in the home are.
Play and exploration are seen as essential to children's growth and development; yet space for exploration and play are currently being reduced.
Cars are seen as bad for the environment; yet children's lives are increasingly determined by them and children are spending more time in them.
Schools are seen as the cornerstone of neighbourhood society; yet increasingly children bypass their local school to attend more distant schools.
Children's intellectual environmental knowledge is increasing; yet their own direct environmental experiences are decreasing.
At the same time that medical and environmental health-related conditions are improving, many children's personal health is worsening as a consequence of social and environmental changes.
As recognition of the need for clean organic environments grows, so too does the prevalence of toxic environments.

Source: adapted from Quortrup (1995)

their needs in their professional realms, and the children themselves. The spaces that children inhabit and the social structures they encounter in those spaces are enormously varied. Some of these are illustrated in Table 1.3. The diversity of family types that children can experience ranges from the so-called standard nuclear family (two to three children and two parents) to children who commute between two homes, which may comprise a single parent in one home and in the other a complex blended family comprising two or more different sets of children coming together at different times. These different family settings take place in very different spatial settings, ranging from, for example, a child living in a city apartment in a deprived but culturally vibrant inner-city neighbourhood, to a child living in a large home in a socially homogeneous gated community.

In the course of a single day a child can access a myriad of different spaces: home, school, park, religious centre, shops, railway station and numerous different street environments. Many children and their families lead complex lives. These are not only complex social lives, but lives that are also spatially complex. The use of space takes place at different scales, from the micro-scale of how children access a room in the house, to the mega-scale where children criss-cross and access different parts of a city. The physical environment for many children has become larger and more complex, but also more fragmented, feared and dispersed. For children attempting to negotiate their way through this world, finding the right path is an ever-changing challenge. Not only does the social environment change, giving different meanings and rules to the spaces that they inhabit, but the spaces themselves change. Professionals trying to provide for children need to be conversant with the changing social rules, expectations and contestations around space, as well as its use, function and the diverse views held regarding for whose use the space is intended. Is the city square a place for active play or quiet conversation? Who are its users and which user groups are welcome and which unwelcome: elderly citizens, young

Table 1.3 Some of the diversity of family types, homes, residential areas and functional places that children experience in the city

Family types	Homes	Spatial character	Functional places
One adult	Small flat/high-rise apartment	Central business district	Home
One adult and siblings	Apartment (low rise)	Inner-city traditional residential	School
Two adults, single child	Townhouse/terraced house	Inner-city, run down	Street
Two adults and siblings	Temporary accommodation (bed and breakfast)	Inner-city gentrified/ regenerated	Health centre
Several adults	Suburban house	Higher-density suburb	Sports/leisure facility
Several adults and siblings	Communal living arrangement	Medium-density suburb	Park
Several adults from different generations	Rented	Low-density suburb	Shops/markets
Extended family as neighbours	Owner occupied	Mixed-housing area	Services (youth club, café, rail station)
Caregiver (non-family environment)	Homeless	Urban periphery	Religious centres
Blended family		Urban fringe/rural	Workplace
Moves between families		Small town or city/ metropolis/megalopolis	Community centre
		Gated community	Open space/green place
		Impermanent (trailer park)	City square/arena, gathering place
			Fun place (theme park, fair ground, open-air theatre)

people, families, homeless people or dog walkers? The square is more than a paved plaza with seats: it has social meanings that influence social inclusion and exclusion; it has written and unwritten rules on its use, and these change over time, even between daytime and evening use.

Thus far, the focus has been on children and the debates around children. Yet, what of children themselves: what is their role; are they to be provided for or are they to be an active participant in the social and physical worlds around them? Within their spatial and social worlds children may be acted upon, but they are not passive reactors. Children themselves shape and influence the city; they colonize streets and create their own spaces in neighbourhoods. They are in a state of continual negotiation over space: where they can go, how they get there, who they go with and what they can do in those spaces. Children need space, access, mobility and independence to develop. As Durkheim observed:

> ... it is plainly evident that one must take into account the child's acutely felt need for movement which, to varying degrees, subsists until adolescence. Any brutal attempt at repression of this tendency would incur the risk of extinguishing the flame which must be kindled. It would choke the keen and joyous impulses of a young life. (cited in Jenks 2005, p27)

Children have a strong need to explore, to encounter, to grow and to engage with the world

Where children's mobility, their inherent needs and their rights to develop are constrained, their well-being will be diminished. But such diminishing is rarely experienced passively. Children have a strong need to explore, to encounter, to grow and to engage with the world. They will attempt to pursue these needs passionately, vigorously and autonomously, using their knowledge, understanding and their social relationships to seek to promote their right to access the spaces that they need. To see this in action you need go no further than the neighbourhood street. As Wheway and Millward (1997) in their study of children living on British housing estates found, children gravitate to the front yard: they want to be where it's at, to see what is going on, to engage with the world beyond. This is why skate parks, tucked away behind buildings out of sight on unwanted bits of land, are of limited value. Children and young people want to practise their skills in public; they want to watch the world go by; they need to access shops, public transport and to socialize as *part of* not *apart from* society. They need to be able to experience at first hand the changes and challenges that are occurring in their own lives and in society at large.

Major Challenges Today

Children's lives are facing huge changes and enormous challenges. Children's lives, as this book shows, have undergone some massive transformations during the last generation, and will undergo more as they grow up. Global challenges are affecting and will continue to affect their lives. Here we mention just a few of these changes. Population growth and decline: children are becoming rarer; some now talk about the 'rarefaction' of children. In this uncommon state, their lives will be less anonymous, more intense and expectations greater. Health is a key issue around children and the population in general. Despite the huge advances in medical science and the provision of health services, children's health and well-being are a great cause for concern. Their physical well-being – notably

Children's lives are facing huge changes and enormous challenges

their weight and fitness – has generated vast amounts of research and dedicated intervention programmes, as has their mental well-being, as ever more children are diagnosed with mental 'illnesses', from attention-deficit/hyperactivity disorder (ADHD) to anxiety and depression-related disorders. The move towards a more egalitarian society remains elusive. The hope of an egalitarian society seems to be no closer now than in the previous generation. More than 80 per cent of the world's population lives in countries where income differentials are widening; thus, inequalities persist and social polarization is increasingly evident, from severely disadvantaged to highly privileged (UNDP, 2007). Even countries generally seen as 'rich' seem to be unable to provide effectively for their children. In a survey of child well-being in rich countries, the US and the UK come bottom of a list of 21 countries placed below Hungary (UNICEF, 2007). What is it that makes The Netherlands take up the leading position and the US the bottom? It is not a matter of resource availability, but something deeper. In looking at economic well-being, paradoxes again emerge. With privilege comes a loss of freedom, loss of independent mobility and increased pressure to succeed at a young age. Finally, there are the major environmental problems, as well as a looming energy crisis, with which children's increasingly car-dependent lives are inextricably bound. Peak oil (the global peak in oil production) may bring benefits as children's lives become more localized, less car dependent and neighbourhood sociability is rejuvenated. Peak oil is but one aspect of the wider challenges. It is not for children to rescue adults from their own environmental folly; but it is for adults to ensure that they pass on the environment in a state that supports children's well-being. Mounting scientific evidence suggests that global climate change will lead to substantial reductions in children's well-being. As well as trying to reduce the negative impacts of global warming, adults now have a responsibility to help prepare our children for a world characterized by uncertainty and immense challenges. Thus, should we be creating spaces and societies that help our children to develop into resilient, capable, adaptable adults, rather than simply preparing them for successful adulthood aimed at a consumerist world?

This Book

Children's lives are simultaneously becoming richer and more deprived

Children's life worlds are multifaceted and are undergoing vast changes reflective of the social and physical transformation of the societies in which they live. Children's lives are simultaneously becoming richer and more deprived. This book explores this conundrum, examining the social and environmental changes affecting children, the key activities, spaces and experiences that children have, and how these can be managed to ensure that children benefit from change. It is organized into three parts. Part I looks at children's worlds, their changing environments and changing lives. Part II focuses on children's activity spaces: home, school, neighbourhood, city centre, service space, cultural space and natural space. Part III examines how to make a difference by creating positive environments for children and addresses key elements in this respect: accessing space; design; and professionals and children working together. The final chapter looks at future challenges in a global context. In the book, the focus is on children in developed countries at all points on the social spectrum from deprived to privileged, recognizing that some children may experience both in different parts or at different times of their lives. The book seeks to embrace childhood as a time of freedom, social engagement and environmental adventure. It looks to the future and the potential impacts of some major looming environmental and social challenges, including peak oil, climate change, environmental degradation, and the social implications of developments such as gated communities, urban regeneration and the move towards higher-density living. Suggestions and directions for redressing the balance in favour of child-supportive environments are included throughout. The book contends that preparing child-supportive environments will have wider benefits, including building more resilient neighbourhoods, cities and societies better able to deal with future societal and environmental challenges, the energy crisis being a critical one. Important in this regard are the voices of children. Children from Australia and New Zealand were asked to contribute to their views on spaces that children in the city use and how these can be improved. Written comments and drawings from children from three schools and one cultural centre have been included. These are included at the end of chapters in Part II. We do not claim that the voices of these groups of children are representative or indicative of wider children's voices; but they are, nonetheless, the voices of children. The points they make are valid and considered and add to the book's value, relevance and richness.

Childhood is a time of freedom, social engagement and environmental adventure

References

Bartlett, S., Hart, R., Satterthwaite, D., de la Barra, X. and Missair, A. (1999) *Cities for Children: Children's Rights, Poverty and Urban Management*, Earthscan, London

Bowlby, J. (1999) *Attachment: Attachment and Loss, Vol 1*, 2nd edition, Basic Books, New York, NY

Boyden, J. with Holden, P. (1991) *Children of the Cities*, Zed Books, London

Chawla, L. (ed) (2002) *Growing Up in an Urbanizing World*, UNESCO and Earthscan, London

deMause, L. (1974) 'The evolution of childhood', in L. deMause (ed) *The History of Childhood*, Psychohistory Press, New York

Donahoo, D. (2007) *Idolising Children*, University of New South Wales Press, Sydney

Driskell, D. (2002) *Creating Better Cities with Children and Youth: A Manual of Participation*, UNESCO Publishing, Paris

Durkheim, E. (1979) 'Childhood', in W. Pickering (ed) *Durkheim: Essays on Morals and Education*, Routledge and Kegan Paul, London, pp27–32

Erikson, E. H. (1993) *Childhood and Society*, W. W. Norton & Co., Inc, New York

Freeman, C. and Aitken-Rose, E. (2005a) 'Future shapers: Children, young people and planning in New Zealand local government', *Environment and Planning* C, vol 23, pp227–246

Freeman, C. and Aitken-Rose, E. (2005b) 'Voices of youth: Planning projects with children and young people in New Zealand local government', *Town Planning Review*, vol 76, no 3, pp287–312

Frones, I. (1995) *Among Peers: On the Meaning of Peers in the Process of Socialization*, Scandinavian University Press, Oslo

Furedi, F. (2008) *Paranoid Parenting: Why Ignoring the Experts May be Best for Your Child*, Continuum, London

Gill, T. (2007) *No Fear: Growing Up in a Risk Averse Society*, Calouste Gulbenkian Foundation, London

Hart, R. (1979) *Children's Experience of Place: A Developmental Study*, Irvington Publishers Inc, New York, NY

Jenks, C. (ed) (2005) *Childhood: Critical Concepts in Sociology, Vol 1*, 2nd edition, Routledge, Oxon

Kytta, M. (2004) 'The extent of children's independent mobility and the number of actualized affordances as criteria for child-friendly environments', *Journal of Environmental Psychology*, vol 24, pp179–198

Leach, P. (2009) *Child Care Today: Getting It Right for Everyone*, Alfred A. Knopf, New York

Liedloff, J. (1977) *The Continuum Concept*, Knopf Publishing Group, New York

Matthews, H., Limb, M. and Taylor, M. (2000) 'The street as thirdspace', in S. Holloway and G. Valentine (eds) *Children's Geographies: Playing, Living, Learning*, Routledge, London, pp63–79

Moore, R. (1986) *Childhood's Domain: Play and Place in Child Development*, Croom Helm, Kent

Piaget, J. (1954) *The Construction of Reality in the Child*, Basic Books, New York

Postman, N. (1994) *The Disappearance of Childhood*, Vintage Books, New York, NY

Quortrup, J. (1995) 'Childhood and modern society: A paradoxical relationship?', in J. Brannan and M. O'Brien (eds) *Childhood and Parenthood: Proceedings of ISA Committee for Family Resource Conference on Children and Families*, Institute of Education, University of London, London, pp189–198

Skenazy, L. (2009) *Free-Range Kids: Giving Our Children the Freedom We Had Without Going Nuts with Worry*, John Wiley & Sons Inc, San Francisco, CA

Spock, B. M. and Needlman, R. (2004) *Dr. Spock's Baby and Child Care*, Pocket Books, Simon & Schuster, New York

Stanley, F., Richardson, S. and Prior, M. (2005) *Children of the Lucky Country? How Australian Society Has Turned Its Back on Children and Why Children Matter*, Pan Macmillan Australia, Sydney

Stearns, P. N. (2006) *Childhood in World History*, Routledge, New York, NY

UNESCO–MOST (United Nations Educational, Scientific and Cultural Organization–Management of Social Transformations) (1994–2003) *Growing Up in Cities Project*, www.unesco.org/most/guic/guicmain.htm

UNDP (United Nations Development Programme) (2007) *Human Development Report 2007/2008*, UNDP, New York

UNICEF (United Nations Children's Fund) (2007) *Child Poverty in Perspective: An Overview of Child Well-Being in Rich Countries*, UNICEF, Innocenti Research Centre, Florence

Valentine, G. (2004) *Public Space and the Culture of Childhood*, Ashgate, Hants, UK

Ward, C. (1978) *Child in the City*, The Architectural Press, London

Wheway, R. and Milward, A. (1997) *Child's Play: Facilitating Play on Housing Estates*, Chartered Institute of Housing, Coventry, UK

2
Same World – Different World

> Childhood is starting to resemble a prison sentence, with children spending almost every moment behind locked doors and alarms, their every movement scheduled, supervised and controlled. Are they at least safer as a result? Probably not. Obesity, diabetes, and the other health problems caused in part by too much time sitting inside are a lot more dangerous than the spectres haunting parental imaginations. (Gardner, 2008, p14)

> Children will always be children and will always find a way to play. (Hughes, 2003, p34)

Varied and Changing Lives

Is the world markedly different for children today? In some ways children today experience their world in much the same way as in previous generations, when they are given the chance. On a recent visit to Central Park in New York City, children could be seen engaging in the same behaviours that their parents probably enjoyed when they were children in the park. Children took every opportunity to climb the rock walls, explore the bushes, roll in the grass and play make-believe games among the statues. Children will always find ways to play. Yet their worlds are changing in complex ways. While parents have always been concerned for their children's safety, current concerns are dominated by fears that may not be justified, leading to constrained worlds in which children are likely to be anxious and unadventurous (Skenazy, 2009).

The most dramatic change in children's lives is their loss of freedom to play

The most dramatic change in children's lives over recent decades has been the loss of children's freedom to engage in unstructured play and to freely explore their own neighbourhoods and cities as they mature. These changes are due to factors such as a growing culture of fear, increased competition among parents to help their children achieve success in a consumerist world, the impacts of changing technologies (particularly mobile phones), and changing household structures and urban forms (e.g. closure of local schools). We can also identify variability in children's lives according to variability in socio-economic status, urban versus rural living, different levels of housing density and different cultures. In this chapter we explore the ways in which children's lives change, or remain the same: across generations, in different cultures and in different urban locations. We also examine how children's views are different from (or the same as) adults'.

Children as Social Actors

Urban societies comprise a variety of social groups with their own lenses through which they see the world. Some of these groups have more power to influence the places in which they live. Hugh Matthews (1995, p456) describes groups close to the centre of urban decision-making as 'insiders', and those whose voices are rarely heard as 'outsiders' – social groups who 'live on the edge of society, within a marginal world whose visions comprise values often at odds with the mainstream'. In many modern Western societies, children's views, wishes, desires and needs are systematically ignored or devalued. While attempts are being made to redress this situation for children of all abilities, urban landscapes are testaments to the contrasts between adults' and children's views of children's worlds.

The way in which adults see children's worlds has changed in many nations, so that children are now more likely to be categorized as outsiders. Adults, including parents, increasingly regard children in public space as out of place. In this way, environments reflect adults' values and usages, with only token spaces left for children. When neoliberalism is the dominant political ideology, children's issues tend to be about keeping public space free of trouble-making young people (White, 1996; Malone and Hasluck, 2002). Urban landscapes reflect the different access that adults and children have to power. Chawla and Malone (2003) noted in their comments on the Growing Up in Cities research initiative that children experience a sense of powerlessness, and that this is consistent across all the cities that were involved in the project. In Chapter 12 we explore the ways in which children can be given voices that enable them to be active participants in the planning and operation of their urban environments.

Same World

Children all over the world have similar views on a quality urban environment

Children all over the world have similar views on what is desirable and what is not in a quality urban environment. The large-scale international project Growing Up in Cities found a remarkable consensus in children's views of what a good city should look like (Chawla, 2002). Children living in a poor neighbourhood in Boca Barracas in Buenos Aires, or children living in Northampton, England, or Trondheim, Norway, identify similar lists of positive and negative indicators in cities (Chawla, 2002, p229). Positive urban qualities include cohesive communities and social integration, green areas, peer-gathering places and freedom of movement. Negative indicators include fear of harassment and crime, social exclusion, pollution and heavy traffic.

Although technology has changed children's experiences, children's priorities have remained reasonably stable over the last several decades. As suggested in the quote by Hughes (2003) at the start of this chapter, children's core business – play – still dominates their interaction with the world: they will always find ways to play, irrespective of the environments designed for them by adults and the restrictions placed upon them. 'Children value the same sorts of places they did decades ago' (Elsley, 2004, p158). They particularly value the informal or wild spaces, places that can be manipulated by children or places that allow children to interact with features of the environment (particularly natural features). The places most valued by children include fields and hills, trees, jumping and climbing places, brooks or frog ponds, hiding places, woods, forts (dens or cubbies) and streets (Matthews and Limb, 1999; Elsley, 2004). Parents today still continue to recognize the importance of outdoor play for children.

Children find affordances that are not immediately obvious to adults

As in previous generations, children's spatial ranges and 'licences' of independent mobility still increase with age, though usually with more restriction than in previous generations (see Chapter 10). Children still find affordances in urban environments that are not immediately obvious to adults. They still have similar fears – strangers, bullying, dogs and traffic. Both parents and children still have these fears (Pain, 2006). Yet, despite the consistency over recent generations in the things and places that children value, there have been significant changes in children's lives: changes in freedom (their independent access to local neighbourhoods), changes in the control and surveillance of their lives, changes in the levels of adult-supervised activity, and changes in the urban environment that limit children's access.

Different Children's Worlds over Recent Decades

For many adults today, a successful childhood is measured not by whether children have access to independent play experiences, but by whether they are adequately prepared to compete with others in a consumerist world. As well as a worldview dominated by competition, several changes or trends in cities and lifestyles over the last 10 to 20 years can be identified as having an influence on children's experiences of urban life (see Table 2.1)

A Growing Ethos of Individualism, and Changes in Housing and Households

In many nations, particularly in suburbs in the US, Australia, Canada and the UK, the ethos of the individual (or the family unit) has come to prominence: 'suburban neighbourhoods everywhere seem to have in common a lack of the mutuality, the permanent but intangible "community spirit" that is characteristic of the urban village' (Knox and Pinch, 2010, p191). This is partly related to the newness of urban communities, as well as the pressures of urban life that make it difficult for residents to find time to interact with their neighbours. When households are trying to maximize the extra-curricular activities for their children, and fit these in with the commute, the shopping and the gym, time stress can result. The impact of this upon children is felt as parents try to compete with other parents to provide their child with the best opportunities, and parents also adopt an individualistic strategy towards protecting their children. This means that children are more likely to be driven to school and to other places (to protect them from traffic danger), and more likely to be kept indoors or under adult supervision.

Parents compete to provide their child with the best opportunities

There have also been marked changes in many new housing developments – smaller blocks and bigger houses. This trend has been particularly marked in Australia over the last 30 to 40 years. From 1985 to 2008, average house sizes in

Table 2.1 Changes in cities and lifestyles affecting children

Changes	Context
A growing ethos of individualism	More privatized lifestyles and home entertainment; more use of private motor vehicles, less use of public space.
Smaller families and households	Trend towards having children later and smaller families.
Increasingly scheduled lives	Increased extra-curricular activities, at least for children in families with regular incomes; less unstructured play and more adult-organized activities for both academic and other activities.
Closure of local shops and schools	Part of a broader trend towards neoliberal ideologies dominating policy-making.
Loss of wild spaces (natural spaces)	Urban development takes over previously unused spaces. Opportunities for children to interact in a naturalized setting are greatly diminishing, and this change has implications for the child's normal healthy development.
Changes in suburban development – smaller blocks and larger houses	A combination of urban consolidation, increasing affluence and an increased focus on the home as a place for living.
Global financial crisis	This may have had some positive impacts, at least for those households where parents are forced to reduce the scheduling of children's lives. It may also have negative consequences as children lose the family home and parents lose jobs and their ability to support their family.
Increased population densities in many cities	More children are now living in high-rise housing, even in cities usually regarded as being low-density cities (e.g. Sydney and Vancouver).
Pressure on children to compete and to succeed	In an increasingly globalized world, cities compete with each other to attract investment, and parents compete with each other to give their children an advantage in a consumerist world.
A replacement of children's independent mobility with adult-dependent mobility	Fewer children now walk or cycle, or use public transport, and more children are taken by car when cars are available. This is being driven by both the increasing distances involved to get to schools, shops and services, as well as a growing culture of fear.
Increased use of technology for surveillance of children	Mobile phones, global positioning systems (GPSs), webcams and closed circuit televisions (CCTVs) are now a part of childhood for many children.
More play indoors rather than outdoors	This is both a response to fears for children's safety in public spaces as well as a consequence of increased availability of indoor entertainment and education possibilities (e.g. computers).
Increased fear or risk culture	This increased fear is not reflected in statistics on children's accidents or crime.
Increased pace of life and time pressure	A reliance on cars exacerbates this increased pace of life, and cities whose transport systems are dominated by private motor vehicles spend more money (and, hence, more time) on transport than those where walking, cycling and public transport are the dominant modes.

Australia increased from 170 to 250 square metres, and average lot size fell from 660 to 630 square metres (Matusik, 2009). Many new estates are also dominated by façades with minimal window frontages (mainly double or triple garages, often with remote-control door openers) and few front porches. The combination of these factors means that children have less outdoor space available to them and more indoor space; streets also have less passive surveillance from the windows of the residences. As a result, many parents perceive the streets to be lonely, deserted and dangerous in terms of 'stranger danger'.

As well as changes to housing, there have been changes in the structure of households and communities. Today, there are fewer multigenerational families in many cities. This means that children are more isolated from adults whom they once had regular contact with, especially grandparents. Added to this is the increasing segregation of people by age, so that moving into nursing homes or retirement villages is seen as the expectation for elderly people. This reduces the passive surveillance by elderly people in the local neighbourhood as children walk or cycle to school, to the shops or to their friends' homes.

Today children are more isolated from adults, especially grandparents

An important demographic trend is towards smaller family sizes, which means that parents with fewer children may be more likely to guard them more fervently. This has led to a phenomenon known as 'helicopter parenting', and even 'helicopter grandparents' who watch children's every move (either by being with them or by using electronic surveillance). Research suggests that this helicopter parenting is widespread throughout the world (Gibbs, 2009).

Access and Mobility

One of the most dramatic changes in children's lives over recent decades (at least in many developed nations) has been the reduction in the independent mobility of children (the freedom of children to explore their neighbourhoods or cities without a supervising adult). In place of this independent mobility, children are now more likely to have adult-dependent mobility, particularly in private motor vehicles. We examine this in detail in Chapter 10, showing how children's journeys are now much less likely to be made by walking, cycling or public transport than they were two or three decades ago. The loss of independent mobility is one of the key indicators of a lack of child friendliness in urban environments.

The reduction of children's independent mobility is related to a host of factors, including the growth of a culture of fear (see below), growth in traffic volumes, and the closure of local shops, services and schools. Research on children's mobility indicates that distance to school is one of the most influential variables affecting children's journey to school or to leisure activities (Fyhri and Hjorthol, 2009). School closures (along with the closure of post offices, health services and local libraries) have led to an increased reliance by children on motorized transport, particularly their parents' cars. Reductions in independent mobility have been particularly marked in the US, the UK and Australia, while countries such as Finland and Germany still have high levels of freedom for children's movement in their neighbourhoods and cities.

Outdoor Play

It may not be surprising to hear that children in many nations have less opportunity for outdoor play than they did a generation ago. What may be surprising is the extent of the decline in children's outdoor play and the changes in the type of play behaviours. A large-scale study in the US surveyed over 800 mothers in a representative nationwide sample (covering cities, small towns, suburbs and rural areas), asking them about their play experiences as children in comparison with their children's experiences: '70 percent of mothers reported playing outdoors every day when they were young, compared with only 31 percent of their children' (Clements, 2004, p72). Not only were today's children much less likely to play outdoors, they were more likely to have passive, inactive lives indoors. Despite the marked decline in children's outdoor play, 93 per cent of mothers believed that such play was important for children's physical and motor skills development, and 75 per cent agreed that outdoor play was important for children's social skill development. Several other studies have found that parents believe that children need and benefit from playing outdoors (Soori and Bhopal, 2002).

The decrease in children's imaginative play is a particular cause for concern

The types of play behaviour engaged in by children had also changed over the last generation in the US study by Clements (2004): 78 per cent of parents reported regularly taking part in imaginative or make-believe games with friends. In contrast, such games were only played regularly by 57 per cent of children. Clements believes that the decrease in children's imaginative play is a particular cause for concern because of the importance of such play for social cognition, language development, and because of opportunities to mimic and interpret adult behaviours. Another type of play that showed a dramatic reduction was the street games using child-initiated rules, which decreased from 85 per cent (mothers) to only 33 per cent for their children.

The only outdoor activities in which children were more likely to be involved than their mothers were organized sporting activities (with a coach or adult present). This was part of a broader trend to engage children in adult-structured activities: 'scheduled play dates, music lessons and after-school youth sports on most days of the week, leaving them with less time to initiate their own activity in school or at home' (Clements, 2004, p73). Free unstructured play is seen as a waste of time by parents primarily motivated by the need to see their children succeed. Parents are now more likely to regard structured activity as 'play', although child development professionals clearly regard structured activity as non-play (Fisher et al, 2008).

A Japanese study by the Benesse Educational Research Centre in Tokyo reported similar findings to the US study by Clements, indicating that indoor play had been 'normalized' for fifth and sixth graders at elementary schools in both metropolitan and rural areas in Japan: 40 per cent of children preferred playing indoors rather than outdoors. The impact of the reductions in children's play behaviours has been seen as deleterious to the development of resilience in children. In fact, one psychologist argues that the result of not letting our children play freely is that we may be raising 'a nation of wimps' (Marano, 2008) who are not developing the sense of place or the resilience that comes from close personal contact with their environment, particularly their natural environment.

Risk

Parents today are concerned with many of the same issues concerning children that attracted their attention a generation or more ago. Importantly, parents aim to protect their children from a range of dangers or risks. Yet, a growing body of commentators and researchers have argued that parental concerns about their children's safety (at least for some types of 'risk') have developed to a level of paranoia (Franklin and Cromby, 2009; Furedi, 2002). While the popular press busies itself disseminating the dangers of the modern world for children, some researchers have suggested a different thesis: 'people worry about an ever increasing range of dangers which in reality are unlikely to happen: a culture of anxiety which is driven by rapid technological development and global insecurity' (Pain, 2006). However, as Pain (2006) argues, sometimes parents' and children's fears may be grounded in actual experience. There are also important gender and racial dimensions to this risk. In Pain's research on children aged 10 to 16 in Gateshead in north-east England, children reported high levels of victimization (including violent crime and harassment) that often matched their levels of fear of crime. Girls had higher levels of fear, but lower levels of victimization, and black and ethnic minority children feared racist victimization. The reported lived experience of children suggests that their fears of public spaces are sometimes not exaggerated. Thus, at least in the area for Pain's research, which included areas that were among the most deprived in the UK, while risks of extreme events (e.g. abduction) may be low, the everyday experience of crime and harassment provides the basis for children's fears. Such experiences may be less typical of less disadvantaged areas, and may also be less evident in other nations such as Germany.

Parents today are still motivated by the desire to be 'good parents'. Yet the concept of a good parent has changed markedly over recent decades in many Western societies. Parents feel increasingly responsible to protect their children from all risks that might interfere with their development into successful consumers. Many parents now see the world as a more dangerous and more competitive place than in previous generations, a world in which children's success, or their survival, requires constant parental surveillance, monitoring, guiding, protection and stimulation. The idea that children 'learn' through play seems to have been largely ignored by many parents, who have decided that children learn best when stimulated through extra-curricular activities. This manifests itself in seemingly ridiculous pressures. For example, in Japan, the parents of toddlers are competing with each other to gain entry into the best kindergartens and preschools. Two- and three-year-old children are enrolled into 'cram schools' to help them pass entrance exams for the best kindergartens. One of Japan's most prestigious elementary schools had over 1300 applicants for just 132 positions in the first grade (WuDunn, 1996).

Are children living in the same world as in previous generations in terms of risks that they are exposed to? Are the dangers to children greater or lower than in previous generations? The two main dangers that parents are concerned about are traffic danger and 'stranger danger' (as well as risks of assault or bullying). While some areas have seen improved traffic safety levels through the introduction of lower speed limits and traffic calming, increasing traffic levels in many urban areas mean that such areas are less safe for children than in previous generations. Although there have been fewer road traffic deaths, there have been increases in minor accidents. One reason for the lower child pedestrian accident deaths is that children have been removed from the streets, rather than the streets becoming safer for children. In the UK, during 1982 to 2002, child pedestrian death rates fell significantly. The most plausible explanation for this downward trend is restricting children's exposure to traffic

Children have been removed from streets, rather than streets becoming safer for children

Figure 2.1 Learning to negotiate traffic is a risky endeavour, but a skill that children have to learn if they are to use the city effectively

Source: Claire Freeman

(Roberts, 1993). As the total volume of motorized traffic increases in urban areas, so too does the level of danger for children and other pedestrians and cyclists (see Figure 2.1).

Skenazy (2009), author of *Free-Range Kids*, argues that we are poor risk assessors because we focus on certain risks and ignore other long-term risks: 'By worrying about the wrong things, we do actual damage to our children, raising them to be anxious and unadventurous or, as [Skenazy] puts it, "hothouse, mama-tied, danger-hallucinating joy extinguishers"' (Gibbs, 2009). Many parents have adopted an individualistic response to traffic danger and stranger danger, choosing to drive children to places and keep them indoors or under supervision. The long-term impacts of this include not only the limiting of children's development, discussed above (as well as their experience of joy and wonder), but also decreased levels of physical activity and increased levels of obesity and type 2 diabetes.

Similar arguments apply to playgrounds, including the widespread adoption of rubber soft-fall surfacing and restrictive rules on children's use of the playgrounds (see Figure 2.2). Playground risks are extremely low in terms of both fatalities and injuries, lower than most sports that children are encouraged to engage in. 'Cost benefit analyses show that residential traffic calming is at least ten times as effective in reducing accident numbers as playground safety surfacing' (Gill, 2007, p29).

In an extreme example of how an individualistic response to risk can dominate parental decisions about keeping their children safe, Skenazy (2009) refers to 'Dear Abby' (an advice column syndicated to 1400 newspapers worldwide), which suggests that parents take a photograph of their children

Figure 2.2 Children are growing up in an increasingly risk-averse society

Source: Claire Freeman

each morning as they leave the house. This way, if the child is kidnapped, an up-to-date photograph can be shown to police! A more collective response to any perceived kidnapping threat would be to increase the level of passive surveillance on the street, which could be achieved by having more people walking in the neighbourhood.

One danger for children that has increased during recent decades is the danger of drowning. In the US, drowning is the second highest cause of death for children (behind car accidents), and three-quarters of pool drownings occur in backyard pools. Yet, the risks of pools for children are poorly understood, even by US parents, who are more concerned about dangers from firearms, despite the lower risks of children being injured or killed by guns compared with pools (Levitt and Dubner, 2005).

Surveillance of Children

Two generations ago, in cities throughout the world, there was an expectation that children would listen to adults and an expectation that adults would keep an eye out for children – a collective responsibility. There were also informal networks of communication in local communities. If a child misbehaved, or had an accident, other adults who knew the neighbours' children would probably contact the parents. Today, while parents are much less likely to be contacted by other adults who are keeping an eye on children, children are likely to be under different forms of surveillance. There is growing use of electronic surveillance – webcams and closed circuit televisions (CCTVs) in homes and schools, as well as tracking devices, including global positioning system (GPS) technology in children's accessories and mobile phones. 'Colleges are installing "Hi, Mom!" webcams in common areas' (Gibbs, 2009). Children in kindergarten are armed with mobile phones in case they need to get help in an emergency. Why? One reason is the constant stream of bad news depicting the dangers of the modern world:

> When webcams distribute images to the audience in the Internet, local gazes are connected with the global community and bodily individuals become intertwined with 'digital individuals' in another sense. (Koskela, 2004, p201)

Today, in most Western countries, most adults and growing numbers of children have a mobile phone, most of which now have cameras. The widespread existence of these camera phones creates a condition reminiscent of surveillance (Koskela, 2004). 'The cell phone means we are always connected to (and attempting to control) our kids' (Skenazy, 2009, p95).

Mobile phones provide a way for parents to keep tabs on their children without being physically present with them. The phone has undoubtedly helped to retain parental control by giving them the opportunity to enter their children's space at any time (Williams and Williams, 2005). Parents expect to be able to communicate with their children at any time, and children are aware of the ability of parents to keep an (electronic) eye on them or, as Williams and Williams (2005) suggest, to invade children's private space. For parents, mobile phones can help to reduce fears of stranger danger. Mobile phones can also allow children to have more freedom, as long as they adhere to rules laid down by parents, such as 'Call me to let me know you've arrived at the sports ground.' Parents and children negotiate a set of rules, including the rule of keeping the phones on, or conventions about texting rather than phoning: 'These phenomena illustrate the degree to which the mobile has quickly become an embedded and vital aspect of relations between parents and their children' (Williams and Williams, 2005).

Children have adopted mobile phones in cities throughout the world over the last decade, and they are now the most popular form of electronic communication for adults and for children. Mobile phones have become a social tool (Campbell, 2005). Yet, mobile phones can also have negative impacts upon children, including through cyber-bullying. Recent research suggests that harassment and bullying by text and voice are becoming significant issues (Pain, 2006, p231). Whether the mobile phone has led to more or less freedom for children is debatable. While children's spatial ranges (and time to roam) may have increased, the extent to which they are out of adults' 'gaze' is called into question when they can be monitored (sometimes with GPS technology) by parents and other adults.

The use of mobile phones raises serious questions about 'trust' and control

The use of mobile phones and other forms of electronic surveillance raises serious questions about 'trust' and control (Rooney, 2010). When parents use electronic surveillance, we believe that we can control the surveillance of our children. This means that we do not have to trust our neighbours to look out for our children, nor do we need to trust our children to look out for themselves. With electronic surveillance we will always be there for them. When this occurs, children 'learn' from their parents' behaviours that they can only trust their parents, and this puts them in a difficult position if they need to call on the help of others. Trusting others is a necessary precursor to building relationships. If we want our children to grow, to learn and to become resilient, we need to learn to trust them to explore the world for themselves. Parents can also experience greater pressures as a consequence of increased use of mobile phones by children. Parents are tied to their children electronically, available at any time, and can be constantly interrupted to fetch and carry children. In this circumstance, children can fail to learn to plan ahead and to make more independent arrangements.

An innovation that can reduce the reliance on mobile phones as a form of surveillance and control is the 'walking school bus' – where children walk to school with other children and supervising adults (see Chapter 10 for more detail). One important benefit of children walking on these buses is that children get to know that there are other adults whom they can trust and whom they can call on if they need to. By walking with other children and other adults, rather than sitting alone in the back seat of their parents' car, children are not only having contact with the local environment, they are also building valuable social connections.

Children get to know that there are other adults whom they can trust

Children's Worlds in Different Contexts: Socio-Economic Status Impacts

While in many ways children living in economically deprived areas are likely to be disadvantaged in comparison to other children, there are complex feedbacks involved. In some ways, children in poorer areas have advantages over children in richer ones. This is particularly the case if children's freedom to explore local environments, and their connection with the local community, are seen as important for children.

In the descriptions of the international research initiative *Growing Up in an Urbanizing World*, Chawla (2002) illustrates the similarities and contrasts between children's lives in different parts of the world. While children value the same things in urban environments wherever they are in the world, they also experience vastly different lives. In some ways, children living in impoverished cities or areas of cities (e.g. Boca-Baraccas in Buenos Aires) may have more enriching lives than children in richer urban areas. In Boca-Baraccas, despite living in deprived environments, children are able to enjoy the cultural richness afforded to them in a dense urban environment. Yet, even within cities in the developed world, we find vast contrasts in children's urban experiences.

In many cities throughout the world, children living in poverty are often disadvantaged in fundamental ways compared to children in more affluent areas. The physical environment of disadvantaged areas is more likely to exhibit indicators of a loss of public order, including graffiti, litter and excrement (Doran and Lees, 2005). Associated with this is the increased incidence of petty crime and vandalism. Children's experience of these spaces is 'likely to be less protected and more vulnerable' (Elsley, 2004, p156). Elsley (2004) describes her research in an urban regeneration area in Edinburgh, Scotland, with children ranging in age from 10 to 14 (the ages when children are given increasing levels of freedom by their parents). Young people in this area reported problems in areas that they played in: unsightly vandalized areas, streams with rubbish, needles in underpasses. These places were places to avoid, not only because of their 'environmental unpleasantness', but also because they were perceived as threatening places. In many disadvantaged areas, children also dislike traffic, as do their parents.

Despite the environmental and social disadvantages for children living in poorer areas of cities, children in lower socio-economic areas may have more independent mobility than children in higher-status areas. Poorer children are often more likely to be allowed to independently explore their local neighbourhoods, and more likely to be able to play outside in public spaces. In extremely deprived neighbourhoods, where crime (and fear of crime) is extremely high, children's freedom is likely to be curtailed. Soori and Bhoopal (2002), in a study involving seven- to nine-year-old children in schools in Newcastle-upon-Tyne, England, found that socio-economically deprived children were more likely than their comparison groups to be allowed to engage in outdoor activities independently. Similarly, Freeman (2010) found children in a disadvantaged public housing area in Dunedin had high levels of independence and social connectivity. These children also did not regard themselves as deprived or as living in a disadvantaged area.

Cultural Differences in Children's Worlds

There are marked international differences in the interaction between adults and children, even between nations with similar levels of economic development. Hillman et al (1990) noted that compared to children in England, German children in public spaces on their own are much more likely to be under the general supervision of adults: 'In parks, on buses and trams, and en route to any destination, children will be observed and "guided" if their behaviour falls short of the standard expected' (Hillman et al, 1990, p84). There is far less fear about 'strangers' in Germany than in England, Australia or the US, as this collective responsibility of adults means that adults can act 'in *loco parentis*'. Thus, a feeling of security for both parents and children is generated by a 'mutual surveillance network' (Hillman et al, 1990, p84).

A culture of individualism developed strongly from the 1980s

The fear of strangers (by both parents and children) is arguably stronger in England, Australia and the US not only because of the lack of this surveillance, but also because of a culture of individualism and privatism that developed strongly from the 1980s. Children (and adults) are increasingly likely to spend their time indoors, or in private spaces, rather than the public spaces of the streets, parks or public transport interchanges.

There may also be cultural differences in the contrasts in children's lives between rural areas and urban areas. While Kytta (1997) found large contrasts in children's freedoms (independent mobility) between rural and urban areas in Finland, studies in the US and England and Wales have found little difference between rural and urban children (Clements, 2004). In Kytta's (1997) study, eight- and nine-year-old children in rural villages and small towns in Finland were much more likely to be given licences to cycle on roads, use buses and go out after dark, than children in a city, though the licences for crossing roads, going to leisure places and coming home from school alone were similar. In

Clements's (2004) research on children's outdoor play, the responses did not vary significantly between rural and urban areas. Research in England and Wales by Smith and Barker (2001) supported this finding, showing that children's access to public space was similar in rural and urban environments.

High-Density Housing and Low-Density Suburbs

Policies to increase residential densities in urban areas have been implemented by city governments throughout the developed world. In many countries this has led to increasing numbers of children living in high-rise buildings, particularly in the inner city. Research in Melbourne indicates the 'relatively rich and complex experiences of children living in central city high rise housing' (Whitzman and Mizrachi, 2009, p58). The research found high rates of children's independent mobility, in comparison to recent studies in developed nations. 85 per cent of parents indicated that their children were permitted to travel without supervision to school, and 59 per cent were allowed to travel to parks or playgrounds by the age of 12. This level of freedom was higher than that found in another Melbourne study of children in suburban areas, where only 59 per cent of girls and 65 per cent of boys aged 8 to 12 were allowed to walk or cycle to school, and 40 per cent of girls and 48 per cent of boys aged 8 to 12 walked or cycled to parks, ovals and playgrounds (Timperio et al, 2004). Differences between children living in public versus private high-rise developments were also evident in the types of places frequented: children from private housing visited a greater variety of places (Mizrachi and Whitzman, 2009): 'While public housing children focused on nearby play spaces, private housing children included private and public play spaces, shops, and landmarks such as the buskers at Flinders Street Station' (Whitzman and Mizrachi, 2009, p58). The lack of backyard play space in higher-density areas means that children are also more likely to seek out spaces outside the home: the street, shops or playgrounds. Hence, children may be more visible in inner-city, high-density areas, despite the lower percentages of children in these areas.

Adults' and Children's Views of the World

Children's ways of seeing the world are rarely acknowledged by adults. While there is a growing international awareness of the importance of children's rights, not much has changed in most cities in terms of children's involvement in planning. Fortunately, there are some examples showing how children can be more involved in the planning of their cities (see Chapter 12).

Children's ways of seeing the world are rarely acknowledged by adults

The concept of affordances is a useful way to understand children's views of the world, and how this is different from adults' views. The affordance of any object, person or place is an indicator of what these may offer to an individual (e.g. child). Thus, for an adult, a set of stairs provides the affordance of getting from one level of a building to another. For children, the same set of stairs can provide a range of affordances, including a place to watch other people, a place to play with their cars or balls, or a place to practise their skating skills. Another example could be the use of fences.

When adults build fences, their main affordance is as boundaries to separate spaces intended for different uses. Children, however, may see them as items in the landscape providing affordances such as climbing and balancing. They can also act as barriers to engagement and exploration, particularly for younger children. In contrast to adults, children are keen to interact with their environment and with each other (see Figure 2.3).

Adults often design spaces without children in mind. Yet children constantly invade those spaces, and use various elements of the environment in ways that enhance their play. Perhaps the most obvious example of a conflict in adults' and children's understanding of affordances is the use of trees. For adults, trees provide shade, food, privacy or noise insulation. For children, trees can provide quiet and secluded places to play, or opportunities for carving, or for climbing and playing hide and seek (see Figure 2.4).

Some spaces within cities are designed with children specifically in mind, and often children enjoy using these spaces. Yet, in many cases, children's favourite places are the most unexpected in adults' eyes (see Box 2.1).

Figure 2.3 Children are keen to interact with their environment and with each other (South Bank, Brisbane, Australia)

Source: Claire Freeman

Figure 2.4 Playing hide and seek in the trees

Source: Claire Freeman

Box 2.1 Affordances in Washington, DC: The Metro

Washington, DC, opened its underground subway system in 1976, and the current 83-station system was completed in 2001. Like the monuments in the city, the Washington Metro has been built on a grand scale. The tunnels, stations and even the subway cars are huge compared to the New York subway. Adults using the Metro may assume that children would find it a dark, scary place.

However, a visitor to Washington from Kansas City told an interesting story. Her children, aged eight and ten, love coming to Washington. This is not because they get a chance to see the White House, or the monuments in the Mall, or the children's exhibitions in the Smithsonian Museum. Their favourite part of Washington is the Metro.

Anyone from Kansas City, with its appalling public transit system, might find an underground subway system something of a novelty. But why do children like it so much? The children from Kansas City liked it for a number of reasons. They 'surfed' the Metro: they played a game with each other to see if they could remain standing without holding on as the Metro cars lurched around the tracks – they liked the sensation of moving and the challenge of staying upright, and spinning around on the poles in the carriages (see Figure 2.5). There is also something magic about disappearing into an underground maze, leaving a street in daylight, then reappearing again in another part of the city. In Washington, this could well seem like another world (to children and to adults). Even getting to the platform is like an adventure – travelling down the long escalators and then into the tunnels. The Metro also provides a (moving) space where children are legitimate users along with adults and people from various backgrounds. Children are not segregated from any other groups on the subway. The Washington Metro provides surprising affordances for children's play.

Figure 2.5 Child playing on a pole on the Washington Metro

Source: Paul Tranter

Reversing the Trends

Many of the changes discussed above that impact negatively upon children also affect adults: invasion of adult space from mobile phones; adults needing to accompany children; the need to work extra hours to buy cars and consumer goods to keep children occupied; and the loss of neighbourliness and social connectivity.

Although the changes in children's lives have mainly involved reduced freedom, increased supervision and reduced access to outdoor or natural environments, there have been several initiatives aimed at reversing these trends, and making spaces in cities more accessible and friendly to children. Among these have been campaigns to reclaim streets for children – for example, the New York City Streets Renaissance Campaign (Kaboom, 2010). Other initiatives include the introduction of lower speed limits and traffic calming (Pucher and Dijkstra, 2003), school gardens and community gardens (Armstrong, 2000), walking school buses and safe routes to school programmes (Kingham and Ussher, 2006; Watson and Dannenberg, 2008), and a growing movement toward 'slow cities' (Honoré, 2004), 'child friendly cities' (Gleeson and Sipe, 2006) and 'healthy cities' (Capon and Blakely, 2008). All of these trends promote slower and more child-friendly modes of living. These and other initiatives are discussed in the following chapters.

References

Armstrong, D. (2000) 'A survey of community gardens in upstate New York: Implications for health promotion and community development', *Health and Place*, vol 6, no 4, pp319–327

Campbell, M. (2005) 'The impact of the mobile phone on young people's social life', *Social Change in the 21st Century*, Centre for Social Change Research, Queensland University of Technology, Queensland

Capon, A. and Blakely, E. (2008) 'Checklist for healthy and sustainable communities', *Journal of Green Building*, vol 3, no 2, pp41–45

Chawla, L. (ed) (2002) *Growing Up in an Urbanizing World*, UNESCO and Earthscan, London

Chawla, L. and Malone, K. (2003) 'Neighbourhood quality in children's eyes', in P. Christensen and M. O'Brien (eds), *Children in the City: Home and Community*, Routledge, London, pp118–141

Clements, R. (2004) 'An investigation of the status of outdoor play', *Contemporary Issues in Early Childhood*, vol 5, no 1, pp68–80

Doran, B. and Lees, B. (2005) 'Investigating the spatiotemporal links between disorder, crime, and the fear of crime', *The Professional Geographer*, vol 57, no 1, pp1–12

Elsley, S. (2004) 'Children's experience of public space', *Children and Society*, vol 18, no 2, pp155–164

Fisher, K., Hirsh-Pasek, K., Golinkoff, R. and Gryfe, S. (2008) 'Conceptual split? Parents' and experts' perceptions of play in the 21st century', *Journal of Applied Developmental Psychology*, vol 29, no 4, pp305–316

Franklin, L. and Cromby, J. (2009) 'Everyday fear: Parenting and childhood in a culture of fear', in L. Franklin and R. Richardson (eds) *The Many Forms of Fear, Horror and Terror*, Inter-Disciplinary Press, Oxford, pp161–174

Freeman, C. (2010) 'Children's neighbourhoods, social centres to "terra incognita"', *Children's Geographies*, vol 8, no 2, pp157–176

Furedi, F. (2002) *Culture of Fear: Risk-Taking and the Morality of Low Expectation*, Continuum, London

Fyhri, A. and Hjorthol, R. (2009) 'Children's independent mobility to school, friends and leisure activities', *Journal of Transport Geography*, vol 17, no 5, pp377–384

Gardner, D. (2008) *Risk: The Science and Politics of Fear*, McClelland & Stewart Ltd, Toronto

Gibbs, N. (2009) 'Can these parents be saved?', *Time*, vol 174, no 21, p52

Gill, T. (2007) *No Fear: Growing Up in a Risk Averse Society*, Calouste Gulbenkian Foundation, London

Gleeson, B. and Sipe, N. (eds) (2006) *Creating Child Friendly Cities: Reinstating Kids in the City*, Routledge, New York, NY

Hillman, M., Adams, J. and Whitelegg, J. (1990) *One False Move: A Study of Children's Independent Mobility*, Policy Studies Institute, London

Honoré, C. (2004) *In Praise of Slow: How a Worldwide Movement Is Challenging the Cult of Speed*, Orion, London

Hughes, F. P. (2003) 'Spontaneous play in the 21st century', in O. Saracho and B. Spodek (eds) *Contemporary Perspectives on Play in Early Childhood Education*, Information Age, Greenwich, CT, pp21–40

Kaboom (2010) *New York City: Streets Renaissance Campaign – Streets as Places to Play*, Kaboom, New York, NY

Kingham, S. and Ussher, S. (2006) 'An assessment of the benefits of the walking school bus in Christchurch, New Zealand', *Transportation Research, Part A: Policy and Practice*, vol 41, no 6, pp502–510

Knox, P. and Pinch, S. (2010) *Urban Social Geography: An Introduction*, 6th edition, Prentice Hall, Harlow

Koskela, H. (2004) 'Webcams, TV shows and mobile phones: Empowering exhibitionism', *Surveillance & Society*, vol 2, no 2/3, pp199–215
Kytta, M. (1997) 'Children's independent mobility in urban, small town, and rural environments', in R. Camstra (ed) *Growing Up in a Changing Urban Landscape*, Royal Van Gorcum, Assen, pp41–52
Levitt, S. and Dubner, S. (2006) *Freakonomics: A Rogue Economist Explores the Hidden Side of Everything*, Penguin Books, Camberwell, Victoria
Malone, K. and Hasluck, L. (2002) 'Australian youth: Aliens in a suburban environment', in L. Chawla (ed) *Growing Up in an Urbanizing World*, UNESCO and Earthscan, London, pp81–109
Marano, H. (2008) *A Nation of Wimps: The High Cost of Invasive Parenting*, Broadway Books, New York, NY
Matthews, H. (1995) 'Living on the edge: Children as outsiders', *Tijdschrift voor Economische en Sociale Geografie* [*Journal of Economic and Social Geography*], vol 86, no 5, pp456–466
Matthews, H. and Limb, M. (1999) 'Defining an agenda for the geography of children: Agenda and prospect', *Progress in Human Geography*, vol 23, no 1, pp61–90
Matusik, M. (2009) 'Small is the new black', *Residential Developer Magazine*, December, available at www.propertyoz.com.au/Article/NewsDetail.aspx?id=2554 (accessed 11 November 2010)
Mizrachi, D. and Whitzman, C. (2009) 'Vertical living kids: Creating supportive environments for children in Melbourne Central City high rises', in *Proceedings of the State of Australian Cities National Conference: City Growth, Sustainability, Vitality and Vulnerability*, Australian Sustainable Cities Research Network, Perth
Pain, R. (2006) 'Paranoid parenting? Rematerializing risk and fear for children', *Social & Cultural Geography*, vol 7, no 2, pp221–243
Pucher, J. and Dijkstra, L. (2003) 'Promoting safe walking and cycling to improve public health: Lessons from The Netherlands and Germany', *American Journal of Public Health*, vol 93, no 9, pp1509–1516
Roberts, I. (1993) 'Why have child pedestrian death rates fallen?', *British Medical Journal*, vol 306, pp1737–1739
Rooney, T. (2010) 'Trusting children: How do surveillance technologies alter a child's experience of trust, risk and responsibility?', *Surveillance & Society*, vol 7, no 3/4, pp344–354
Skenazy, L. (2009) *Free-Range Kids: Giving Our Children the Freedom We Had Without Going Nuts with Worry*, John Wiley & Sons Inc, San Francisco, CA
Smith, F. and Barker, J. (2001) 'Commodifying the countryside: The impact of out of school care on rural landscapes of children's play', *Area*, vol 33, no 2, pp169–176
Soori, H. and Bhopal, R. (2002) 'Parental permission for children's independent outdoor activities: Implications for injury prevention', *The European Journal of Public Health*, vol 12, no 2, pp104–109
Timperio, A., Crawford, D., Telford, A. and Salmon, J. (2004) 'Perceptions about the local neighborhood and walking and cycling among children', *Preventive Medicine*, vol 38, no 1, pp39–47
Watson, M. and Dannenberg, M. D. (2008) 'Investment in safe routes to school projects: Public health benefits for the larger community', *Preventing Chronic Disease*, vol 5, no 3, www.ncbi.nlm.nih.gov/pmc/articles/PMC2483559/pdf/PCD2483553A2483590.pdf
White, R. (1996) 'No-go in the fortress city: Young people, inequality and space', *Urban Policy and Research*, vol 14, no 1, pp37–50
Whitzman, C. and Mizrachi, D. (2009) *Final Report: Vertical Living Kids: Creating Supportive High Rise Environments for Children in Melbourne*, Australia, VicHealth, Melbourne, p71
Williams, S. and Williams, L. (2005) 'Space invaders: The negotiation of teenage boundaries through the mobile phone', *The Sociological Review*, vol 53, no 2, pp314–331
WuDunn, S. (1996) 'In Japan, even toddlers feel the pressure to excel', *New York Times*, 23 January

Part II

Activity Spaces

3
Home

> Walking into Green Acres, you immediately sense that you have walked into an oasis – traffic noise left behind, negative urban distractions out of sight, children playing and running on the grass, adults puttering on plant-filled balconies. Innumerable social and physical clues communicate to visitors and residents alike a sense of home and neighbourhood. This is a place people are proud of, a place children will remember in later years with nostalgia and affection, a place that feels 'good'. (Cooper Marcus and Sarkissian, 1986, p1)

Introduction

Children spend more of their life in their home than anywhere else. Even if they go to school or day care and have active social lives with friends and family, they return to their family home. Yet, the role, function, design and use of the family home have been little studied. It seems implausible that so little attention should have been paid to such a central space in children's lives; yet such is the case. There has, indeed, been strong focus given in media and in research to the family and matters such as different types of family and the factors making a family 'functional' or 'dysfunctional'. Property developers give much attention to the notion of 'the family home'; but again little attention is given to how families actually relate to space, to what families really want or need, rather than what developers imagine they want and need. This chapter addresses this deficit and explores the concept of the home: its design, function, use and association as a space lived in by children and their families. Homes comprise two important facets: the physical building, and the social constructions that are afforded to the building and its use. This chapter examines both of these factors. It begins by looking at the housing needs of children and the fact that these are often not met. It investigates how homes meet – or fail to meet – the changing needs of children and their families, focusing on children, families and social change, and how use of the home responds to societal developments.

The role, function, design and use of the family home have been little studied

Variations on a Home Theme

The notion of homes as externally constructed spaces, in whose shape or form their inhabitants have little say, is a relatively recent innovation. Historically, residents were much more intimately connected with their construction. In many parts of the world, homes are still built by those who will use them, or are handed down through the family and repaired, modified and extended as required and as family circumstances change. The notion of mass-produced housing, with large groups of similarly designed buildings to which families must conform, is at odds with the more incremental type of housing development that characterized and, in some places, still characterizes how houses have traditionally developed. Houses historically have reflected the culture, land, beliefs, aesthetics and practical circumstances of the society in which they are built. Hence, their enormous variation from the tall, thin canal houses of Amsterdam, the predominantly wooden Japanese homes with their internally flexible room spaces, to the courtyard housing of hotter climates found in Spain and Italy, as well as Arabic and Chinese courtyard homes. More recently built houses may still retain some of these traditional features, but invariably reflect ever more universal international housing styles that only tangentially reflect the cultural, spatial or other needs of the families who are to dwell in them. In some places there is little option other than the mass-produced apartments or suburban estates that proliferate in cities around the world.

In the face of this standardization of housing provision and the difficulties of reversing bad housing design and development, it is imperative to understand housing from the perspectives of its current and intended inhabitants, and to build (and build better) according to their needs. The homes that people like are geographically and culturally determined. In the UK, for example, the two-storey terrace or semi-detached house with private front and back garden remains popular with British families compared to high-rise apartments, which have never been well liked. Yet, high-rise apartments are the predominant housing type and work in some of the more highly urbanized Asian countries such as Singapore and Hong Kong. For high-rise apartments to work for children and their families, they need to be supported by the availability of high-quality accessible public space (see Figure 3.1). Many Europeans happily live in apartments in their compact city environments, having access to comparatively limited private open space, whereas most Australians and North Americans seem to make little compromise on the detached dwelling as their lifestyle choice, even if this exacerbates suburban sprawl. Yet, even in these countries there are places where extremely high-density living is the norm. Manhattan is perhaps the best-known example.

Homes reflect dominant societal ideologies

Homes reflect dominant societal ideologies. This can be seen in the mass housing characteristic of Eastern bloc states: uniform apartments with little acknowledgement of family and individual needs. It can also be seen in the dominance of the detached suburban dwellings characteristic of aspirations towards home and landownership as part of the Australian or American dream, even if these aspirations are at the expense of increasingly long commutes and longer working hours. Within this housing diversity it is imperative to ask: what are the needs of children and their families? How well can these be met by the existing housing provision and how can, and should, a new domestic landscape emerge?

Children's Housing Needs

The real needs of children are generic and transcend housing type, style and often culture. Professed needs such as a personal bedroom or private backyard can deflect attention away from more fundamental needs. In 1986, Cooper Marcus and Sarkissian wrote their seminal book *Housing as if People*

Figure 3.1 These homes were not designed for children, but the reality is that children live in them: In high-rise Hong Kong, the need to provide accessible outdoor communal play and activity space provision is imperative

Source: Michelle Thompson-Fawcett

Mattered. It identifies key physical needs for good housing for children, including:

- safe outdoor play area (not needing constant parental supervision);
- safe from traffic and pollution;
- natural spaces (places with flexible, malleable materials);
- private open space that is linked to communal open space;
- communal spaces for adults and children to meet each other;
- private play spaces;
- good management and maintenance regimes;
- house identity and variety in buildings;
- street linkage and access to a wider environment that encourages independence.

While Cooper Marcus and Sarkissian's book is not a recent publication, it reflects good thinking on the topic that transcends the short timeframes commonly used by house builders and governments today. Indeed, a recent

Figure 3.2 Well-planned homes with good open relationships with the street encourage social contact and play between children and young people

Note: These children regularly play and spend time together after school and on weekends.

Source: Claire Freeman

publication by City of Portland, Oregon (2007), *Principles of Child Friendly Housing*, begins by referencing Cooper Marcus and Sarkissian, reinforcing their work's pivotal position in housing literature. Portland's good housing criteria also focus strongly on external links to outdoor space and to the wider community (see Figure 3.2).

The needs of children in relation to housing remain remarkably consistent across time and space, and typically include safety, security, a sense of belonging, a mixture of family and personal space and family/communal space, play space, and a healthy environment (see Table 3.1). Common to all principles of good housing is a sense of children as social beings – the need for children and their families to interact, to socialize and to meet friends and neighbours without effort. There should be a link between private spaces and public spaces within the house so that children have a place to retreat to and have time out and a place where they know they can meet others in the family. Children and families also need this link to be reflected at the broader level so that families can have space they acknowledge as private, but can also access communal spaces for support and social activity.

Play features heavily in the design needs of children. Children need places for quiet play, but also for interactive and active outdoor play: play that can take place safely and unsupervised. Central to all other needs is the need for a sense of belonging and security, for children to feel attached to their home and for the home to act as a safe base from which they can gradually go out, explore, develop independence and experience the city and society beyond. An important element is that of a semi-private space, or a buffer zone between the private space of the home and the public space of the neighbourhood. This space allows young children to transition from play in the home to play in the neighbourhood.

Table 3.1 Children's housing needs

Good housing provides the following	Bad housing contributes to the following
Sense of belonging	Poor physical health
Stability	Poor mental health, stress and depression
Safety and security	Insecurity
Healthy environment	Family breakdown
Independence	Learning and behavioural difficulties
Private space	Under-achievement
Family space	Low aspirations
Places in which to build relationships	Increased domestic strife
Access to other children and adults	Higher accident rates
Play space – active and quiet play	Higher rates of child abuse
Sense of identity	Poor school attendance and achievement
Semi-private space	Lower long-term life chances

Lack of attention to these positive attributes results in a range of problems for families and children, as is indicated in the list of bad housing factors in Table 3.1. The focus in Table 3.1 is on generic factors related to well-being – not on a set of design guidelines – indicating that what matters are fundamental principles, and it is with these in mind that homes should be built.

When Housing Goes Wrong

The needs of children in housing are most apparent when 'housing goes wrong'. Poor housing and poverty usually go together. What is especially disconcerting is that in many ways problems in this regard are increasing rather than decreasing. Child poverty has risen in 17 out of the 24 Organisation for Economic Co-operation and Development (OECD) countries for which data is available. The impacts of bad housing upon children are described by Harker (2006, p7), author of a housing study in the UK:

The needs of children in housing are most apparent when 'housing goes wrong'

> Children living in poor or overcrowded conditions are more likely to have respiratory problems, to be at risk of infections and have mental health problems. Housing that is in poor condition or overcrowded also threatens children's safety. The impact on children's development is both immediate and long term; growing up in poor or overcrowded housing has been found to have a lasting impact on a child's health and well-being throughout their life.

Another study by Habitat for Humanity on housing in the US estimated 30 million households there faced housing problems. This included 6.1 million living in overcrowded conditions, with 5.1 million American households facing 'worst-case housing needs', a figure that includes some 3.6 million children

(Habitat for Humanity International, 2009). In the UK it was estimated that in 2005, 1.6 million children lived in bad housing or were homeless. Housing conditions tend to be worst in some of the larger cities, and while London features internationally as sixth in a league table of the world's richest cities, conditions for its children are not so good. Over 650,000 children in London live in poverty, 41 per cent of the total number of children, with 16,240 households being identified as homeless (Mayor of London, 2007). A further 46,000 households with children were in temporary accommodation, including the highly unsuitable bed-and-breakfast accommodation. This is where children and their families are provided with a single room in a hotel. However, the word hotel can be misleading as these are usually poor-quality, overcrowded establishments that are not designed for self-catering, so families find it difficult to undertake normal activities such as cooking, washing and socializing. Studies have shown a link between poor housing and increased child abuse, with overcrowding and frequent moves being identified as significant contributory factors (Sidebotham et al, 2002).

For homeless children and children in temporary accommodation, the problems associated with bad housing are exacerbated, overcrowding is worse and frequent moves lead to a lack of stability. There are often problems attending school, and physical and mental health problems intensify. Housing problems can be particularly acute for certain population groups and tend to be worse for ethnic minorities, and refugee and lone parent families (e.g. 1 in 12 Bangladeshi households in London are in housing need). While problems in London may be particularly difficult, it does not have a monopoly on poor housing. It was estimated that throughout England there were 544,000 overcrowded homes, most of which would include children (Mayor of London, 2007).

A key feature in providing adequate housing is affordability, and again in many countries housing affordability is decreasing. As a general guide, housing is deemed affordable if it costs less than 30 per cent of the household's disposable income. The 2008 *New Zealand Social Report* estimated that between 1988 and 1997, the proportion spending more than 30 per cent on housing rose from 11 to 25 per cent of households, before levelling off at 24 per cent between 1998 and 2001, rising again in 2007 to 26 per cent (Ministry of Social Development, 2008). Overall statistics can mask wide discrepancies in access to good housing, especially for indigenous populations in countries such as New Zealand, the US, Australia and Canada, and for ethnic and other usually less privileged populations internationally. For the 14.8 million US households that make US$10,000 or less per year, a year's rent costs about 70 per cent of their annual income (Habitat for Humanity International, 2009). Poor housing is not inevitable. As the UNICEF (2007) report shows: 'there is no obvious relationship between levels of child well-being and a country's gross national product', the commonly used indicator of economic well-being. The UNICEF (2007) report *Child Poverty in Perspective* placed Poland and the Czech Republic higher on its child well-being table than France, Austria, the US and the UK, even though these nations are usually thought of as wealthier countries. Poor housing is prevalent in most countries, but least prevalent in the Nordic countries, where inequality differentials are less evident.

The Child Poverty in Perspective *report placed Poland and the Czech Republic higher on child well-being than France, Austria, the US and the UK*

Homes for 'Changing' Families?

In considering homes for families, the challenge is to recognize the diverse needs of families and the wide range of cultural, social and physical situations within which families live. Families are diverse, ranging from a single adult and child to a mixture of adults comprising different generations, differently aged children and possibly children and adults with whom the children may or may not be related. These variations are indicated in the family types that were identified in Table 1.3. As families have become more mobile within and between countries, traditional family structures, supports and ways of living break down and are replaced by others. Aitken (1998), in his book *Family Fantasies and Community Space*, identifies the 'friends as family' where friend networks perform many of the social and support functions previously associated with families and their relations. Families have also become more fluid as partnerships change, and families may themselves change over time as children spend time in different homes with different adults. The term 'blended family' has been used to describe some of these more complex family types.

The challenge is to recognize the wide range of cultural, social and physical situations within which families live

There are some key changes occurring in societal functioning that are influencing children's home relationships. Four significant ones are identified here. First, there are more households without children. This means that more children will live in neighbourhoods where there will be fewer children, even in areas traditionally seen as dominated by 'family housing'. Associated with this trend is a lower tolerance of children as part of the domestic landscape, reflected in lower acceptance of children's play in the streets, and in behaviours such as refusal by neighbours to return children's balls that have gone astray. Second, as work hours and commuting times increase, children are living in settings where more adults (including parents) are absent from the neighbourhood for longer periods of the working day. In turn, this means that children have access to fewer potential carers and thus are themselves more likely to be absent from the home area for longer. Third, homes increasingly house fewer adults and are commonly designed to house the 'nuclear family' of two adults and two or three children, rather than the extended family. Fewer adults being available to undertake care of children again reduces the amount of caring that can be done in the child's own home. Where homes do house extended families, frequently these are ethnic minority families such as Bangladeshi families in the UK, Pasifika families in New Zealand, or Chinese families in California, US. Typical 'nuclear family' homes are often too small and ill designed for multiple generations, resulting in overcrowding. Families and their children can face increased exposure to the effects of poor housing, as identified in Table 3.1.

The fourth key change is that parental divorce and separation means that many children inhabit more than one home as they move between homes and between parents. Some homes are close but often can be in different cities, distant parts of the country or even different countries. The number of children living in two homes appears to be increasing. Data on the number of children in this position is hard to establish as most arrangements are informal and fall outside of any statistical information-gathering process. However, some particularly useful studies have been undertaken in the UK (Smart et al, 2001) and Australia (Smyth et al, 2008) on families' post-separation arrangements. In 2003, 1 million Australian children were estimated to be living with only one natural parent. Furthermore, in December 2006 it was estimated that 9.5 per cent of non-resident parents had children in their care for at least 30 per cent of nights a year, compared with only 4.4 per cent in June 1999 (Ellison, cited in Smyth and Moloney, 2008). The range of post-separation arrangements varies, but generally falls into the following categories:

- equal (or near equal) shared care;
- weekly or fortnightly arrangements involving overnight stays;
- daytime-only contact;
- holiday-only contact (typically because of large distances between parents' households);
- sporadic or intermittent contact;
- little or no parent–child contact (Smyth, 2005).

In the first four arrangements children will be involved in some movement between homes. The 2006 Australian Family Law Amendment (Shared Parental Responsibility) Act states that:

> Courts with family law jurisdiction in Australia now have a responsibility to consider making orders for the children to spend equal or else substantial or significant periods of time with each parent where such arrangements are in children's best interest and reasonably practicable. (Smyth and Moloney, 2008, p7)

Shared care is being increasingly promoted as the ideal and the norm by courts of law (Gilmore, 2006). In the main, research has focused on children's post-separation experiences with reference to their emotional attachment to their parents; little attention has been given to the implications for children who move between homes with regard to attachment to place, the home and the neighbourhood, as well as access to friends and school, or to the practical implications of living in two homes. In the shared care scenario, children are engaged in a process of recurrent adjustment to living in different places, with different rules, associations and possessions, and it is unlikely that friend and neighbour networks will be equally accessible in both places. Some of the difficulties around the continual change are evident in the child interviews reported by Neale and Flowerdew (2007, p33). One of the children in their study describes her arrangement:

> *Rachel (aged 17):* They both arranged their lives around us ... They both sued for custody, 'cause both my parents wanted to have an equal part in my growing up ... I was going back and forth between them ... I was with my Dad on Mondays and Tuesdays, my Mum on Wednesdays and Thursdays, and then we would alternate weekends. And then it was mad on Sundays.

Children can have the status of temporary residents where the adults are permanent residents

The children have the status of temporary residents in a place where the adults are permanent residents. Few adults would be expected to live lives in a continual process of motion. Yet, for children this is increasingly being seen as the 'ideal' and reinforced in courts through shared care arrangements. While there can be undoubted benefits to shared care, the implications for children and the child's ability to have a say in determining their place of residence needs to be studied and prioritized (see Rhoades and Boyd, 2004; McIntosh and Chisholm, 2008).

Another example of children having two homes can be seen in the experiences of (usually) immigrant children whose families have moved to a new country. Their families often desire retention of links with the family 'home' and culture, and children can spend extended periods in their parents' home country. The conceptions of home evidenced by children extend to more abstract notions of home, and home as sites of belonging where children perform 'multiple and intersecting identities' in the different spheres of their lives (Ni Laoire et al, 2010). Children form complex attachments to different physical places and to social (usually family) connections associated with the notion of 'home', reflective of what can be mobile and transnational lives. This sense of multiple belongings was clearly evident in Mand's (2010) study of Bangladeshi children in London, where the children saw both London and Bangladesh as 'home'. Of the 55 children aged nine to ten in the study, most had been to Sylhet, Bangladesh, two or three times. Home was commonly defined by reference to family as being 'where your family is', with the children showing strong 'home' connections to both countries. As well as moving physically between 'homes' mostly during holidays, they also moved between very different cultural and living environments. These visits are highly valued by children and their families as an essential way of maintaining family ties. In other cases children can be sent 'home' to ensure their 'safety', a situation that is often associated with the parents' uncertain immigration status, a common issue for immigrants in countries such as the US and European countries with high immigrant populations, such as France, Germany and the UK. These children can experience difficulties educationally if their absence is prolonged and can also experience difficulties returning to the original family home where their own or their family's residence status may be insecure. Children's experience of home is diverse. Homes reflect external societal changes which impact upon children and their families and to which they must adjust. Another major area of change is how much time children spend at home.

Homes reflect societal changes which impact upon children

Changes in the Family Home: More or Less Time at Home?

How homes are used has changed markedly in the space of just one generation. The biggest change is perhaps the amount of time that children spend in the home. This time in the home increases as children's outdoor play decreases in line with reduced independent mobility. But it decreases as children are cared for more in formal settings, such as after-school care or crèches, rather than being cared for at home. The reduction in independent mobility and associated cut in children's free-range play and exploration has been well researched by authors such as Jones and Cunningham (1999), Tranter (2006), Mackett et al (2007), Freeman and Quigg (2009), Ridgewell et al (2009) and others. It is a trend that seems well established in the US, Australasia and much of Europe. The corollary of this trend is that children spend longer periods indoors. In her study of Amsterdam, Lia Karsten (2005) found that during the 1950s and 1960s, the street was primarily a child space used for play, but is now increasingly an adult space with car drivers as its key users. For private home space, the reverse is true as adult space becomes increasingly a child space. The

How homes are used has changed markedly in the space of just one generation

reasons for this are changes in family, changes in the outdoor environment and changing societal views. Family size in Amsterdam has decreased from 3.75 in 1950 to 1.98 in 2000, resulting in more indoor space being available (Karsten, 2005). Streets have become the preserve of traffic, making them less conducive to play and less safe places, generally. Socially, outdoor play seems to have become less tolerated, and in Karsten's (2005) study of three Amsterdam streets has diminished substantially. She did find rates of street play varied, with higher rates of outdoor play occurring among some of the larger immigrant families who are more likely to have less indoor space.

Backyards have also changed – from functional spaces to 'designer spaces'

The characteristics of backyards have also changed over recent decades in many nations from functional spaces (e.g. vegetable gardens) or flexible spaces, to 'designer spaces'. A key theme here is the popularity of backyard makeover television programmes, such as *Backyard Blitz* in Australia. In these programmes, backyards where children might have been free to dig, make dens or create their own play spaces are often replaced with designer backyards. These are typically spaces with 'neat' landscaping, 'architectural plants', hard surfacing and 'entertainment spaces'. Little space is left for children, and, if so, it is often a prefabricated children's play set. The emphasis seems to be on increasing the market value of the home, rather than increasing the value of the backyard for children's play.

As parents work longer hours and extend commuting times, their lives and the lives of their children become more pressured, and time spent away from the home increases. Internationally, the trend in working hours is towards overall fewer hours. However, the trend is uneven between and within countries. Those in higher professional jobs and those in multiple low-paid, part-time jobs tend to be worst off in terms of hours worked. The number of hours worked needs to be considered in association with time spent commuting to work, which can add two or more hours to the time spent away from the family home. Canada's working hours remained reasonably stable between 1976 and 2008; but work hours for couples did increase from 57.6 to 64.8 per week. However, whereas in 1976 it was estimated that one third of couples had two earners, in 2008 this had increased to three-quarters of all couples, reflecting a larger number of families with two working parents (Marshall, 2009). Where more parents work, childcare is delegated to people and places external to the family home, before- and after-school care, holiday programmes, holiday camps, child carers and a range of other care options.

As parents work longer hours, time spent away from home increases

Access to statistics on childcare arrangements is difficult as arrangements are often informal and may often be 'illegal', as with informal paid childcare arrangements – not declared for tax, or where children are left to 'self-care'. One source that is available is the US Census Bureau report *Who's Minding the Kids*, which reports that 63 per cent of under fives are in a regular care arrangement, with 35 per cent being in non-relative care – and of these, only 3.7 per cent are cared for in the family home (Overturf Johnson, 2005). Thus, 31 per cent of children spend considerable parts of their day away from the family home in care, the commonest facilities being organized day-care facilities. Even when cared for by relatives, such care will usually be away from the family home. Older children also experience much of their time outside the home. In addition to time spent at school, some 53 per cent of American 5- to 14-year-olds are in a childcare arrangement on a regular basis (Overturf Johnson, 2005).

What of children who are home on their own either because their parents are at work or through the child's own choosing? In a number of countries there has been explicit legislation placing limits on children being left without adult supervision, even in the family home. This means that although children could go home after school, for them to do so can expose parents to prosecution if there is no adult at home. In New Zealand it is illegal to leave a child under the age of 14 without adult supervision except for a very short time. UK law does not set a minimum age at which children can be left alone; however, it is an offence to leave a child alone when doing so puts them at risk. The guide produced by the National Society for the Prevention of Cruelty to Children in the UK suggests children under 12 should only be alone for very short periods and no child should be alone overnight until age 16 (NSPCC, 2009). In the US a few states give 10 as the minimum age for a child to be left unsupervised, but 12 is more general. As in the UK, the principle is generally one of the child not being put at risk. There is, though, a significant disconnect between what the law states and children's real experiences. In few countries is there any before- or after-school care for older children, and few children want to be 'cared' for in this way. Costs and lack of availability of care also exclude many parents from using these options. The US Census Bureau's report estimates that 6.1 million children aged 5 to 14 care for themselves on a regular basis, on average, for 6.3 hours per week. Most of these children are in the older age group (Overturf Johnson, 2005). Thus, children's right to be present in their own homes is subject not only to parental decision-making, but to external (often legal) provisions and requirements. Regulation of the child's right to be at home is a peculiarly Western phenomenon and one that, while generally well intended, can further restrict children's opportunities to have self-directed time at home and can fail to recognize children's own competence.

Regulation of the child's right to be at home is a peculiarly Western phenomenon

Changes in the Family Home: Use of the Home and Home 'Rules'

There has again been remarkably little written on children's home experiences, how they use space, what spaces they commonly inhabit, their own views on these spaces, and what makes a good home space for children. Although it is clear what constitutes a bad home for children – physically and emotionally – once basic housing conditions of space, warmth, cleanliness and adequate furnishings are met, there is little understanding of what else matters. We know little about the answers to simple questions such as do children want or need a separate play space, or do they prefer to be and play in the communal meeting space; what furnishings matter to them; what is their relationship with the kitchen and eating areas; do they prefer entry by front or back doors; and what sort of sleeping space do they prefer? Not much is known about how the use of space is decided, negotiated and enforced. One aspect of space use that has been the focus of attention is that of children's relationship to, and use of, space for technology – notably, televisions and, latterly, computers.

Technology space

As many families have become smaller and space is more available, it becomes possible for family members to define spaces in the home for specific activities – office, study, computer space, sewing, rumpus room – in a way that is not possible with smaller or more crowded homes. However, in Japan, children's dedicated homework space is considered a very important part of the household regardless of its size (see Figure 3.3). Two activity spaces increasingly common in households internationally are TV and computer activity spaces. There has been much debate by educationalists, psychologists, social workers and other child professionals about TV. In addition to concerns over how much TV should be viewed and whether TV influences behaviour, concern has been expressed about having TV as the focal point of the family space. More recent debate has been about TVs in children's bedrooms, and whether this is removing children from the social family space and what the implications of this might be. Certainly, the trend is for increasing availability of TVs in bedrooms for children of all ages. Kotler et al (2001) estimate that 60 per cent of American adolescents and 30 per cent of preschoolers have access to a bedroom TV. An Australian study on children's TV viewing found, by comparison, that 19 per cent of children had a bedroom TV (van Zutphen et al, 2007). Associated with the perceived increase in TV viewing generally, and particularly bedroom technology, has been rising concern over whether this increases children's inactivity levels and, thus, affects their general health through increased body mass index (BMI), an indicator of healthy or unhealthy weights, as well as their developing social skills. The Australian study found overweight or obese children were more likely to watch more TV and have a TV in their bedroom than healthy weight children (van Zutphen et al, 2007). There is no doubt that more children now have technological media such as TVs, DVDs and computers in their bedrooms and spend increasing amounts of time engaging with these.

Figure 3.3 When Japanese children start primary school at the age of six they receive two presents to mark the start of their serious study period: a desk and a leather-made *Trandoseru* (school bag)

Note: The desk is specifically designed for growing children in the primary school years (6 to 12 years of age), with adjustable heights for the desktop, a desk light, a book shelf, drawers, etc. This dedicated study space is an important part of the family home.

Source: Masato Kadobayashi

Whether the problems caused by these trends are outweighed by the benefits (e.g. access to information) is unclear.

Computers, like TVs, have the power to change how space is used in the home and can influence family social relations. Holloway and Valentines' book *Cyberkids* (2003) provides a discussion on issues around children's and families' computer use and what 'cyberspace' means in the home. Their research shows the wide range of ways in which computer use is controlled, negotiated and allocated spatially and temporally in homes, and how its use can both encourage and discourage social interaction. Most interestingly, the study found that anxieties around computers eroding social interaction and outdoor play were largely misplaced. They found that computer use itself is often a social activity, its use tending to replace 'doing nothing' time or TV time rather than outdoor or activity time. In their study, most children preferred to be outside when weather and light allowed. Of the 54 per cent of non-school time spent at home, only 5 per cent was spent on computers compared to 25 per cent watching TV. Furthermore, children's computer use often took place in tandem with socializing with family or watching TV. In some families, use of the computer enhanced family time as various family members joined in games or online searches.

Anxieties around computers eroding social interaction and play were largely misplaced

Home rules

The home is essentially an adult space; its design, purchase, location, internal furnishing and allocation of use is essentially adult determined. This is particularly evident in the promulgation of rules around the house. Some rules are common to homes across diverse cultures, such as you eat at the communal eating place – possibly a table; in cold climates, outside doors are kept shut; noise is reduced at night; babies may sleep with their parents, but older children sleep in a different space; visitors enter by the front door; and beds are tidied when you get up. Other rules are culturally determined, such as shoes are not to be worn indoors in Japan, or candlelight is preferred to harsh light for socializing in Denmark (known as *hygge*, a concept roughly translating to cosiness, comfort and quietness, and is associated with the company of family and close friends). There are also a host of rules specific to each home and to the objects in the home that the child learns; some will be common to other homes, while others are specific to the child's home. A truly eye-opening book is *Home Rules* in which the authors (Wood and Beck, 1994) study the rules around just one room: the lounge in Wood's home. The book clearly portrays the myriad of rules that govern not just the lives of the author's children, but the lives of all children. In any house each house/room/object will have its own rules. For example, rules for a back door may include:

The home is an adult space – evident in the rules of the house

- Keep the back door closed (to keep the heat in and the flies out).
- Keep the back door open (to let the heat out).
- Don't slam the door.
- Don't leave the door unlocked when you go out.
- Don't shut the door if the dog is outside (or else the dog can't get in).
- Don't put sticky fingers on the door handle.
- Wipe your feet on the mat before you come in through the door.

Or the TV:

- Don't watch TV in that chair (too close).
- Don't turn it too loud.
- Don't eat dinner in front of the TV.
- Don't talk when Mum's watching the news.
- Don't take drinks near the TV.
- Don't bump the TV.
- Don't surf the channels.
- Make sure you turn it off at the power mains at night.
- Put the remote back where it belongs.

There are rules around even the simplest of actions in the home, such as where the plate is put on the table; where shoes may or may not go; what objects (books, ornaments) adults can touch but children can't; which lights can be used and at what time of day. Some rules are universally understood, such as don't fill the bath to overflowing, while others are generally understood, such as don't leave towels on the floor after use; others are more specific (in this family each person has their own towel). Children learn to negotiate this rule maze. Where children are in shared care, they learn the rules of two homes. Children also learn to negotiate the rules present in their friends', neighbours' and relatives' homes.

The home is spatially indicative of adult power. Unlike in the public or social realm, there are few limitations on adult power. For some children, home can be a particularly oppressive environment in which they have few rights. They may not experience the right to a conflict-free environment, the right to self-determination of when and which household facilities can be used, the right to privacy, or the right to be heard and have a say. Children may, in turn, challenge some home rules and, in time, add their own, such as 'Only I can use the green cup', 'Knock before you enter my room' and 'Don't use my bike without asking my permission first.' The home is thus a contradictory environment, one in which children spend a considerable amount of time, one where they can feel safest, but also one where they can experience least support, respect and have few rights.

Homes Matter

The home is the most important space that a child will inhabit

The home, then, is the most immediate and important space that a child will inhabit; it is also arguably the least researched and understood space, and one where the child's experiences can be extremely varied. Homes are where children can feel both most and least secure, where they can be well cared for in congenial, healthy, spacious conditions, or alternatively live in unhealthy, crowded and insecure conditions. The design of homes can support or frustrate children's well-being, and while some design features are essential, such as access to indoor and outdoor play space, the form that this takes can be diverse. Outdoor play space can range from a large private backyard to a small communal courtyard or a safe street adjoining the house. Unfortunately, many homes, though intended for families, pay scant attention to children's real needs: economics, political priorities and the views of builders, architects and

Figure 3.4
Unfortunately, many new homes are being built that are alienating in their design, as indicated in these homes, which preclude any opportunity to interact socially with neighbours or the street environment (Auckland, New Zealand)

Source: Mark McGuire

Table 3.2 Children's views on what makes a great home

<table>
<tr><td>When a child is sick there is always somewhere warm for them to go.</td><td>Getting to have warm rooms and big backyards, as well as good-sized rooms.</td></tr>
<tr><td>I like my house because I have my own room and my room is really big.</td><td>TV, food, couches, beds, drawers, bathrooms, and a garage for skating.</td></tr>
<tr><td>There is lots of space to move around in.</td><td>All the space and thinking space.</td></tr>
<tr><td>My house is a nice house because that is where my family is and I am safe.</td><td>Bigger rooms and for each house to have infinite internet access.</td></tr>
<tr><td>You get your own room and toys.</td><td>A good house has personal space, a big backyard and big rooms.</td></tr>
<tr><td>My house is great because the kitchen and bathroom have a lot of space and it is very clean.</td><td>A house should have lots of space, especially bedrooms and probably a big garden.</td></tr>
<tr><td>Houses are very big and roomy, but the best part is your own room where you have space.</td><td>It's a place where you feel safe.</td></tr>
<tr><td>Big and grassy.</td><td>My house is fun.</td></tr>
<tr><td>My house is a good size to play.</td><td>You have heaps of space.</td></tr>
<tr><td>What makes my house a great house is my bedroom because I have a TV to watch and a PS2 to play.</td><td>I wish there was a lot of space in my kitchen because our kitchen is like a thin passageway.</td></tr>
<tr><td>Heaps of room to skate in for kids and other people.</td><td>Lots of space and grass to play makes a great house.</td></tr>
<tr><td>A great house is a loving family and you get along with your brothers and sisters.</td><td>The house is clean and warm and spending family time.</td></tr>
</table>

developers override considerations of children's needs. While there have been vast improvements in housing conditions, generally, too many children still live in homes that compromise their physical and emotional development, and too many new homes are being built that will prolong this trend (see Figure 3.4).

When children in Dunedin identified the things that make a house a great home for children, important themes were space (inside and outside, as well as personal space) and a feeling of safety and warmth. Some children valued a private bedroom, particularly with access to their own television and PlayStation. The garden and grass were also important (see Table 3.2). Many of the themes identified by Cooper Marcus and Sarkissian (1986) in *Housing as if People Mattered* (outlined earlier in the chapter) are reflected in the themes identified by children.

References

Aitken, S. (1998) *Family Fantasies and Community Space*, Rutgers University Press, New Brunswick, NJ

City of Portland, Oregon (2007) *Principles of Child Friendly Housing*, Bureau of Planning, City of Portland, OR

Cooper Marcus, C. and Sarkissian, W. (1986) *Housing as if People Mattered: Site Design Guidelines for the Planning of Medium-Density Family Housing*, University of California Press, Berkley, CA

Freeman, C. and Quigg, R. (2009) 'Commuting Lives: Children's mobility and energy use', *Journal of Environmental Planning and Management*, vol 52, no 3, pp393–412

Gilmore, S. (2006) 'Contact/shared residence and child well-being: Research evidence and its implications for legal decision-making', *International Journal of Law, Policy and the Family*, vol 20, p3, pp344–365

Habitat for Humanity International (2009) 'Affordable Housing Statistics', available at www.habitat.org/how/stats.aspx, accessed 1 September 2009

Harker, L. (2006) *Chance of a Lifetime: The Impact of Bad Housing on Children's Lives*, Calouste Gulbenkian Foundation and Shelter, London

Holloway, S. and Valentine, G. (2003) *Cyberkids: Children in the Information Age*, Routledge Farmer, London

Jones, M. and Cunningham, C. (1999) 'The expanding worlds of middle childhood', in E. Kenworthy Teather (ed) *Embodied Geographies: Spaces, Bodies and Rites of Passage*, Routledge, London, pp27–42

Karsten, L. (2005) 'It all used to be better? Different generations on continuity and change in urban children's daily use of space', *Children's Geographies*, vol 3, no 3, pp275–290

Kotler, J. A. Wright, J. C. and Huston, A. C. (2001) 'TV use in families with children', in J. Bryant and J. A. Bryant (eds) *TV and the American Family*, Laurence Erlbaum Associates Inc, New Jersey

Mackett, R., Brown, B., Yi Gong, Kitazawa, K. and Paskins, J. (2007) 'Children's independent movement in the local environment', *Built Environment*, vol 33, no 4, pp454–468

Mand, K. (2010) '"I've got two houses. One in Bangladesh and one in London ... everybody has": Home, locality and belonging(s)', *Childhood*, vol 17, no 2, pp273–287

Marshall, K. (2009) 'The family work week', *Perspectives – Stats Canada*, Catalogue no 75-001-X, pp5–13, www.statcan.gc.ca

Mayor of London (2007) *The State of London's Children Report*, Greater London Authority, London, September

McIntosh, J. and Chisholm, R. (2008) 'Cautionary notes on the shared care of children in conflicted parental separation', *Journal of Family Studies*, vol 14, no 1, pp37–52

Ministry of Social Development (2008) *The Social Report: Indicators of Social Well-Being in New Zealand*, NZ Government, Wellington

Neale, B. and Flowerdew, J. (2007) 'New structures, new agency: The dynamics of child–parent relationships after divorce', *International Journal of Children's Rights*, vol 15, pp25–42

Ni Laoire, C., Carpena-Mendez, F., Tyrrell, N. and White, A. (2010) 'Introduction: Childhood and migration – mobilities, homes and belongings', *Childhood*, vol 17, no 2, pp155–162

NSPCC (National Society for the Prevention of Cruelty to Children) (2009) *Home Alone: Your Guide to Keeping Your Child Safe*, www.nspcc.org.uk, accessed 1 September 2009

Overturf Johnson, J. (2005) *Who's Minding the Kids? Child Care Arrangements: Winter 2002*, US Census Bureau, Department of Commerce Economics and Statistics Administration, www.census.gov/prod/2005pubs/p70-101.pdf

Rhoades, H. and Boyd, S. B. (2004) 'Reforming custody laws: A comparative study', *International Journal of Law, Policy and the Family*, vol 18, no 2, pp119–146

Ridgewell, C., Sipe, N. and Buchanan, N. (2009) 'School travel modes: Factors influencing parental choice in four Brisbane schools', *Urban Policy and Research*, vol 27, no 1, pp43–57

Sidebotham, P., Heron, J. and Golding, J. (ALSPAC Study Team) (2002) 'Child maltreatment in the "children of the nineties": Deprivation, class, and social networks in a UK sample', *Child Abuse & Neglect*, vol 26, pp1243–1259

Smart, C., Neale, B. and Wade, A. (2001) *The Changing Experience of Childhood*, Polity Press, Cambridge, UK

Smyth, B. (ed) (2004) *Parent–Child Contact and Post-Separation Parenting Arrangements*, Research Report No 9, Australian Institute of Family Studies, Melbourne

Smyth, B. (2005) 'Parent–child contact in Australia: Exploring five different post-separation patterns of parenting', *International Journal of Law, Policy and the Family*, vol 19, no 1, pp1–22

Smyth, B. and Moloney, L. (2008) 'Changes in patterns of parenting over time: A brief review', *Journal of Family Studies*, vol 14, no 1, pp7–22

Smyth, B., Weston, R., Moloney, L., Richardson, N. and Temple, J. (2008) 'Changes in patterns of post-separation parenting over time: Recent Australian data', *Journal of Family Studies*, vol 14, no 1, pp23–26

Tranter, P. (2006) 'Overcoming social traps: A key to child friendly cities', in B. Gleeson and N. Snipe (eds) *Creating Child Friendly Cities: Reinstating Kids in the City*, Routledge, Oxon, pp121–135

UNICEF (United Nations Children's Fund) (2007) *Child Poverty in Perspective: An Overview of Child Well Being in Rich Countries*, UNICEF, Innocenti Research Centre, Florence

van Zutphen, M., Bell, A. C., Kremer, P. J. and Swinburn, B. A. (2007) 'Association between the family environment and television viewing in Australian children', *Journal of Paediatrics and Child Health*, vol 43, pp458–463

Wood, D. and Beck, R. J. (1994) *Home Rules*, John Hopkins University Press, Baltimore, MD

4
School

> With the exception of the rare field trip, children only go outside to play. We teach them, by inference, that real learning happens inside and is composed of something other than their own natural observation. (Smith, 1992, p61)

School Grounds as Environments for Play and Learning

Schools became an important part of the urban landscape for children during the 19th century. Harsh conditions for children during the Industrial Revolution led to the emergence of a children's rights movement, and universal schooling became an accepted practice in industrialized nations. An important conceptualization of childhood developed where children were seen as conceptually different from adults and as 'adults in training'. This notion of adults in training is still dominant in most Western societies, and schools are increasingly seen as places where children should be given the best opportunities to develop under a formal and structured education system, to subsequently become successful adults in a consumerist world. As we will discuss in the final chapter, such a focus may be misguided, and may also be depriving children of important developmental experiences, as well as restricting the achievement of their right to play.

This chapter explores the role of the school as a space for children in modern cities, focusing on school grounds, rather than classrooms, and on the informal (hidden) curriculum of schools rather than the formal curriculum designed by education authorities. We examine changes in children's access to unstructured play opportunities in their own school grounds, and how these changes reproduce reductions in play opportunities in their own neighbourhoods and cities. The chapter then explores the complex and interrelated reasons for the decline in children's play opportunities in their schools, and also explains the importance of the informal curriculum in terms of what has been referred to as 'environmental learning'. We outline interventions designed to encourage children's physical activity and play in schools. As well as school ground greening projects, we examine opportunities that arise from introducing unstructured loose materials into school grounds, an idea growing out of a concept first described as the 'theory of loose parts' (Nicholson, 1971). The chapter concludes with a discussion of the possibility that the underlying model of schools may need to be questioned if we wish to maximize the value of schools for our children (Steen, 2003).

Opportunities arise from loose materials in school grounds

Schools as Microcosms of Changes in Neighbourhoods and Cities

School is one of the few places where children engage in active outdoor play

Children's freedom to playfully explore their own neighbourhoods and cities has been declining in many developed countries over the last few decades (see Chapter 10). For many children, the school is one of the few remaining places where they can engage in active outdoor play with other children. School can also provide a place of safety for children from challenged homes, or children who are experiencing difficulties in other parts of their lives. In primary (elementary) schools in Australia, Canada, the UK and the US, up to one quarter of the school day can be spent outdoors in the school ground. Thus, schools provide not only the space, but also the time for active outdoor play. Yet, even at the scale of school grounds, we can see a loss of children's freedom to engage in play. There has been a radical transformation in many schools, where children now have less space and time for play, particularly outdoor play and play that involves contact with nature. Children also have to endure changes in school culture, where outdoor play is restricted by a plethora of rules and regulations. This has been occurring despite the growing awareness of the importance of environmental learning in school grounds (Tranter and Malone, 2004).

Another reason why children's play is restricted is the obsession of adults with keeping their children, their backyards and their school grounds 'neat and tidy'. Thus, children are discouraged from playing in the dirt (or mud) in many schools, and are strongly discouraged from leaving the school ground in an untidy state. Yet, children – left to their own devices – exhibit a strong desire to manipulate their environment in their play, digging holes and channels, making dams, making dens (huts or cubbies), and moving materials around. While this may well give a school ground an appearance of being untidy, perhaps what children really need is to be allowed the freedom for a bit of creative untidiness (see Figure 4.1).

The loss of children's access to play opportunities in their own school grounds has been documented in research in Australia, the UK and the US. In Australia, some disturbing trends have reduced children's opportunities for creative and diverse play. The most serious of these include the reduction in the time given for recess (lunch and other recess periods); the amalgamation of schools in the name of greater economic efficiency; the enforcement of 'out of bounds' areas in parts of the school ground where children would benefit most from play in 'green spaces' (Tranter and Malone, 2008); and the removal of play equipment and the implementation of restrictive rules about children's use of school grounds that force teachers into a policing role (Evans, 1995, 1997, 1998). In the US, by the 1980s, school districts were limiting children's recess time and imposing rules on children to eliminate risks. Some schools went to the extreme of posting signs saying 'No Running on the Playground', and some schools chose not to provide playgrounds at all (Frost, 2007). Other schools prohibit any play before or after school as the ground is not 'supervised'. In English primary schools, between 1990 and 1996, the lunchtime break had been reduced by 26 per cent (Gill, 2007, p66). Another change in school policy is the trend towards structured exercise classes as a way of mandating children's physical activity (as discussed below).

Figure 4.1 An ideal school ground from a child's perspective? Creative untidiness in Orana School ground, Canberra

Source: Paul Tranter

Box 4.1 Two school grounds in Seattle, Washington

School grounds communicate messages to children about school culture that can influence their values and behaviour, as illustrated in the following story.

In Seattle, Washington, two schools stand side by side: one public and one private. The public school lies at the bottom of a slope, with its asphalt play area confined by a chain-link fence. The private school, at the top of the slope, contains diverse plantings and an intimate courtyard with quiet seating areas. What messages do these environments offer to the children who use them each day? And what learning potentials are afforded in them? Susan Janko, a professor of education at the University of Washington, had her students observe these environments. The messages they found were clear. One student noted: the outdoor area of the private school prepares its students for navigating an ivy-league campus, and the public school's outdoor area prepares its students for navigating a prison yard (Johnson, 2000, p1).

In any city, anywhere in the world, we could picture schools with vastly contrasting school grounds (even if not standing side by side, as in the example above). Although the differences between such schools may seem obvious, a possibly more important dimension of the school environment is the school policy and cultures on the use of the school grounds. Even a school with a visually impressive natural environment, with trees, shrubs, walkways and gardens, will be of little value to children if these areas are classed as out of bounds. A stimulating school ground is an important asset for children; but it is of limited use if school policies are restrictive of children's play.

School grounds transmit messages about the school and its place in the world

These trends represent important changes in either or both the culture and the landscape of schools. Schools are symbolic places, as well as places for experience. School grounds transmit messages about the school and its place in the world, and they also transmit messages to children about how their childhood is constructed and about children's place in the school. Box 4.1 provides a vivid example of the symbolic value of school grounds from an American perspective.

National Playground Standards and Risk Anxiety

Reasons for changes in children's access to play in their own school grounds are complex. One factor identified in the US, also likely to have had an impact in other developed nations, relates to national playground standards, which were developed and rapidly applied from the early 1980s, leading to the replacement of much traditional playground play equipment with standardized play equipment that met safety specifications. Such safety specifications were limiting in terms of preventing children from taking the risks that they arguably need in order to learn and develop (Bundy et al, 2009). The safety regulations were amplified by lawyers who brought legal judgements against schools, parks and childcare centres. The common reaction to this by school administrators and teachers, as well as by the designers and installers of equipment, was to eliminate risk almost entirely from playgrounds and school grounds. Not only were school grounds denuded of exciting play equipment that could lead to accident or injury, but restrictive rules led to the loss of children's freedom to use what was there.

The removal of play equipment and restrictions on play at school may be related to a general societal trend where 'risk anxiety' permeates modern life. This risk anxiety is compounded by the conceptualization of children as vulnerable and in need of protection from the risks that modern society has created. This has led to the progressive restriction in children's opportunities for independent play. More and more play activities are organized and supervised by adults, and held indoors (Isenberg, 2002; Tranter, 2006).

These changes in children's lives are motivated by a laudable desire to keep children safe. However, adults focus on a limited range of risks: physical injury while playing, traffic danger and stranger danger. Other dangers that result from restricting children's freedom to play seem to be largely ignored by parents, teachers and policy-makers (Stephenson, 2003). Some researchers have identified the risks of children becoming afraid to use their bodies actively (Steinsvik, 2004). These longer-term risks include obesity and related health issues, and restricted development (physical, intellectual, emotional and social) (Hart, 2002).

One way of understanding the emphasis on protecting children from all risks is to examine the tension that has developed between physical activity and safety. While both are important, the dominance of 'risk anxiety' in modern societies has led many parents and teachers to feel that any risk is too much risk, and that unstructured physical activity is 'unsafe' (Thomson, 2003). Consequently, school boards try to remove all risk (and exciting and challenging activities) from the school grounds, even though this restricts children's development and their understanding and appreciation of their world. This situation has been described as 'surplus safety' (Buchanan, 1999). School grounds are now more likely to reflect what adults believe to be risk free, rather than what children would desire in terms of a challenging and stimulating play environment.

Studies in primary schools in England and Ireland indicate that children are being restricted in their play behaviours by excessive concerns about safety – protection from injury while playing (Thomson, 2007). An English researcher argued: 'teachers should realise the fear of risk is out of proportion to the reality of the number of serious accidents that occur in the playground' (Thomson, 2003, p57).

Play England (2008) in the UK sensibly points out that children should be protected from 'unacceptable risk of life-threatening or permanently disabling injury in play'. The council also recognizes that there will always be some risk of injury when children play, and that this risk is a critically important aspect of children's development:

> Exposure to the risk of injury, and experience of actual minor injuries, is a universal part of childhood. Such experiences have a positive role in child development. When children sustain or witness injuries, they gain direct experience of the consequences of their actions and choices. (Play England, 2008)

Teachers have now emulated parents' excessive concern with safety, fearing serious repercussions from parents if anything goes amiss (even from scraped knees or muddy clothes) (Thomson, 2003). Consequently, teachers place tight controls on school playgrounds – and children become more sedentary (Thomson, 2007). 'Risk anxiety' has invaded the school playground. A fear of litigation pervades school grounds, where parents suing school boards after children are injured can be of more concern to school authorities than the risks to children themselves (Thomson, 2002).

'Risk anxiety' has invaded the school playground

Apart from the dangers associated with inactivity through surplus safety, the strategy of removing all risk from school grounds may be futile in terms of keeping children safe due to the operation of 'risk compensation'. This is the process whereby children (and adults) compensate for any change in the level of risk that they are exposed to (Adams, 1995; Wilde, 1998). Assuming that all individuals have a desired level of risk, if the environment becomes more dangerous, individuals take more care. Conversely, if the environment is made safer, individuals take more risks. Thus, as adults attempt to remove all risks from a school ground, this may simply encourage riskier behaviours so that children's 'target level' of risk is achieved (Wilde, 1998).

If the school ground is perceived to be boring, children may compensate by engaging in a range of activities that increase their level of challenge, excitement and risk (see Figure 4.2). Such activities can include fighting and bullying, climbing on school buildings, taking their risks in other environments where there is less adult surveillance (e.g. in a local wild space or in a local street), or by using play equipment in ways that were not intended by their adult designers so that extra challenges are created (e.g. climbing onto the top of a swing set and jumping from the top bar) (Stephenson, 2003). When children become

Figure 4.2 Risk compensation: Children making play equipment more exciting by taking risks not predicted by the designers of the play equipment

Source: Freeman

bored and 'add their own risks and challenges', safety can be compromised (Elliot, 2008, p12). In contrast, where schools allow children to engage in play behaviours and locations that could be seen as risky (e.g. building their own play huts and dens, or digging holes), there are typically few occurrences of aggressive behaviour between children (Tranter and Malone, 2004). Similar findings have been shown for children in Canada (Herrington and Lesmeister, 2006, p77).

In some countries, there is a greater acceptance of risk-taking for children (Mayes and Chittenden, 2001). Some Norwegian and Japanese school grounds are designed to encourage children to take risks and to take on challenges in their play. In some Japanese kindergartens, play equipment with high wooden beams and walls creates challenges for children with different levels of ability. In Norwegian schools and kindergartens, while obvious high level risks are removed, elements that stimulate risk and challenge are preserved because it is recognized that 'children need to learn how to assess danger' (Steinsvik, 2004, p10).

Lack of Understanding of the Value of Play in School Grounds

Another likely reason for restrictions on children's play is the lack of understanding of the value of play for children, particularly play that involves contact with nature. Play is an essential, but often undervalued, aspect of childhood. It has a fundamental role in children's development (Isenberg, 2002) and learning (Singer et al, 2006). Play is how children experience their world, and discover and learn about themselves and others. It involves taking risks, problem-solving and interacting with people and places. Learning occurs through trial-and-error play events. Play has been defined as 'the spontaneous activity in which children engage to amuse and to occupy themselves' (Burdette and Whitaker, 2005, p46). Play is pleasurable and engaging for children, and can be apparently without purpose (Timmons et al, 2007, p124).

Learning occurs through trial-and-error play events

As well as a general undervaluing of play for its own intrinsic value (i.e. it is fun), there is a lack of understanding of how play provides educational opportunities, particularly in the area of social skills and environmental learning. The learning that occurs through play is often considered peripheral to that which occurs in the classroom (Evans, 1997, p14). There is a widely held belief among teachers that children only need recess to let off steam before they start their schoolwork again. This belief is grounded in the 'surplus energy theory'. Several researchers (Evans and Pellegrini, 1997; Lambert, 1999) believe this theory to be seriously flawed. Such researchers argue that children's play in outdoor environments in schools can make an important contribution to children's development and education.

It is still widely believed, by parents and by educators, that classroom-based intellectual activity provides the most effective way of developing children's cognitive ability. Yet, a wide body of research suggests that play is also vital for this type of development (Bergen, 2002). If children are deprived of play experiences, their development in problem-solving, maths, science and reading is likely to be reduced. In other words, the level of play in the school grounds during recess supports the development of cognitive ability in the classroom. Yet, despite the value of play for academic performance, perhaps ironically, the emphasis on academic performance has resulted in a loss of children's play opportunities at school.

Perhaps the most important reason for the reductions in children's freedom to play within their own school grounds concerns the growing emphasis on academic performance at schools. Education policy-makers in many nations have concentrated on student performance in the standard school curriculum as a measure of success for children. In the US, one manifestation of this was the No Child Left Behind Act, which required testing of children's academic skills, and which led to punitive measures being taken against low-performing schools. An unfortunate response was that 'recess was abandoned by a growing number of schools to make more time for teaching the tests' (Frost, 2007, p17). Not only has children's recess time been eroded, but they also have less time for independent play before and after school due to parents engaging them in extra-curricular activities.

In response to the range of factors that have deprived children of valuable play experiences at their schools, several initiatives are now emerging that show potential in reclaiming play space and play time for children at schools. These include school ground greening projects, forest schools, interventions in school grounds designed to enhance children's play, and some schools starting to reduce the academic workload for children (Honoré, 2009).

Emphasis on academic performance (and seeing children as adults in training – giving them the best chance of success in a competitive world) has also led to changes in the way in which schools are chosen by parents. Parents who can afford to make a choice now are more likely to send their child to the 'best' school, even if that means driving their children past a school that their children could walk to (this issue will be addressed again in Chapter 10). But schools are not just about academic education. As well as the formal education provided at schools, there is also an informal curriculum (or hidden curriculum) where the concept of environmental learning becomes important.

Environmental Learning in School Grounds

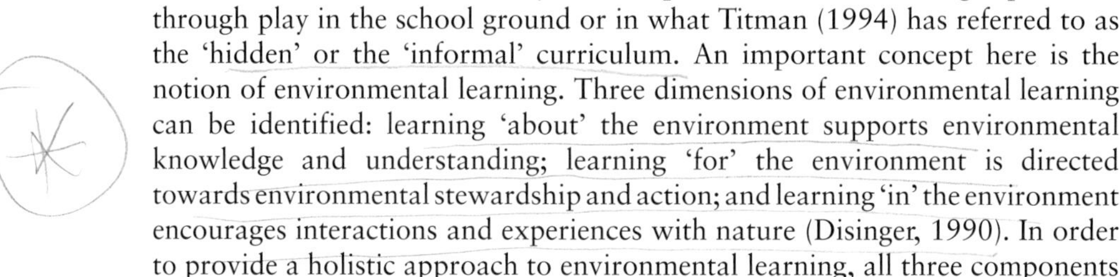

As will be clear from the discussion above, schools not only provide formal education in the classroom. They can also provide valuable learning experiences through play in the school ground or in what Titman (1994) has referred to as the 'hidden' or the 'informal' curriculum. An important concept here is the notion of environmental learning. Three dimensions of environmental learning can be identified: learning 'about' the environment supports environmental knowledge and understanding; learning 'for' the environment is directed towards environmental stewardship and action; and learning 'in' the environment encourages interactions and experiences with nature (Disinger, 1990). In order to provide a holistic approach to environmental learning, all three components should be available through teacher-directed and unguided experiences throughout children's schooling.

School grounds are typically regarded as areas for sport and play, and have often been overlooked for their role in 'education'. Indeed, in most schools in the UK, the US, Canada and Australia, school grounds include various sporting facilities, such as ovals for football, baseball or cricket, or basketball/netball courts. However, school grounds can also provide access to a rich resource of real and natural experiences: exploration of living and non-living things, food webs and life-cycling. Stimulating and diverse school grounds increase the value of play and the range of play behaviours; bland and boring play spaces limit behaviour, restrict opportunities for imaginative play and contact with nature, and enhance behavioural problems such as fighting and bullying (Evans, 1997; Moore and Wong, 1997). The school ground in this way offers a range of 'affordances'.

There is a developmental dimension to the environment

The affordances of an environment are the elements that it offers or provides for the user (Kytta, 2004). As the individual's psychological and physical characteristics change developmentally, the resources that the environment offers also change. For example, an environment that offers the opportunity for the child to climb or hide underneath elements, or contains features that are malleable or can be manipulated, is perceived, used and transformed in different ways at different stages of the child's development. Thus, there is a developmental dimension to the environment, just as there is for the individual child. The utilization of the outdoor environment increases with the child's age, alongside their cognitive, affective and behavioural capacities; the environment should, therefore, be designed to facilitate, support and encourage this developmental growth (Uzzell, 1990). Children play more (and more creatively) in playgrounds with high degrees of challenge, novelty and complexity (Fjortoft and Sageie, 2000). Environments that can be manipulated (e.g. by digging, building and moving objects) provide more environmental affordances and, hence, more play behaviours (Moore and Wong, 1997).

Ideally, school grounds should provide a diversity of environments. Wendy Titman (1994, p58) identified four elements that children looked for in school grounds:

1 *a place for doing*, which offered opportunities for physical activities, for 'doing' all kinds of things, and which recognized their needs to extend themselves, develop new skills, to find challenges and to take risks;
2 *a place for thinking*, which provided intellectual stimulation, things which they could discover and study and learn about, by themselves and with friends, and which allowed them to explore, discover and understand more about the world in which they live;
3 *a place for feeling*, which presented colour, beauty and interest, which engendered a sense of ownership and pride and belonging, in which they could be small without feeling vulnerable, and where they could care for the place and people in it and feel cared for themselves;
4 *a place for being*, which allowed them to 'be' themselves, which recognized their individuality, their need to have a private persona in a public place, for privacy, for being alone with friends, for being quiet outside of the noisy classroom, and for being a child.

A number of elements have been argued to be important in a high-quality school ground. These include water features; possibilities for children to choose their own play activities and to create their own play places; access to nature (trees, ponds, shrubs, flowers, long grass, insects and animals); fields to play on; places and features to sit on, lean against or hide in; and an unstructured environment that can be manipulated, including loose materials for children to play with (Moore, 1986; Titman, 1992; White and Stoecklin, 1998; Fjortoft and Sageie, 2000). Such elements are likely to support high levels of environmental learning in schools.

Several action research projects have demonstrated the potential positive impacts on children's play of improving the design of school grounds. In other words, changing the spaces in which children play at their schools can produce significant improvements in children's play, as well as their contact with nature – or environmental learning. One carefully documented school ground design project involved a school ground in Berkeley, California (see Box 4.2).

School ground features can influence the social hierarchy among children

School ground features also influence the social hierarchy among children. Barbour (1999) compared school playgrounds with very different opportunities for children: either emphasizing physical exercise play or providing a diverse range of play environments. Grounds that promote exercise play encourage gross motor activity as the means for children to interact with peers. Such school grounds favour children with high levels of physical ability; but children with low physical competence are 'constrained by their reluctance or inability to participate' (Barbour, 1999, p94). However, diverse play equipment and materials can support children with low physical competence, and some features (e.g. construction opportunities) encourage cooperative rather than competitive play. In such playgrounds, the dominant children are the imaginative children, rather than the stronger or more physically capable children.

Box 4.2 Washington Elementary Schoolyard: Stimulating environmental learning

A remarkable example of transforming an uninspiring school ground into a stimulating space for environmental learning, and a valued community open space, was the project undertaken by Robin Moore and Herb Wong at the Washington Elementary School in Berkeley, California. This project, which began in 1971, is thoroughly documented in the book *Natural Learning: The Life History of an Environmental Schoolyard* (Moore and Wong, 1997). It demonstrates the impact upon children's play of redesigning the school grounds with the active engagement of children in the process. A large section of the asphalt grounds was reconfigured into natural features, such as woodland, gardens and ponds. This coincided with changes in the social relationships between children and their play behaviours. In the transformed natural areas of the school ground, there were more positive relationships with other children and children exhibited more creative and imaginative play and learning activity. The change in school ground design also encouraged teachers to utilize the new space as an outdoor classroom, reinforcing and connecting with children's play experiences as part of the formal curriculum. Children took on the new role of being knowledge generators rather than just knowledge consumers.

The epilogue to this story, while disappointing from the viewpoint of the school and local community, provides valuable insights into the importance of treating children as capable social actors and as active citizens in their own right. The school was renovated during 1995 to 1996, and many of the natural play spaces were removed, including 30 shade trees, which were replaced with grass and basketball courts. The school authorities who oversaw the renovation argued that they were increasing the space for ball play for older children, mainly boys. However, 'this was achieved at the cost of the diversity of experience and comforting microclimate for all children. This same issue of ball play versus diversity of choice was resolved in favor of diversity when the Yard was first created' (Moore and Wong, 1997, p251).

Moore and Wong, while obviously disappointed at the lack of consultation with either the children or the community, can see some positives in the process. While the physical spaces for play no longer exist in the school ground, the school yard 'lives on through the … students who benefited from growing up with the experiential richness of nature each school day' (Moore and Wong, 1997, p252). There are also important lessons here for involving children in decisions about their schools. Until adults (particularly those on school boards and education authorities) see children as active citizens, rather than adults in training, it may not be possible to create the spaces needed for environmental learning in schools.

School Ground Interventions: Enhancing Children's Physical Activity and Play

Because of the time that children spend at school and the restrictions on children's freedoms in many neighbourhoods, schools are seen as an important area for promoting physical activity among children. Several school ground interventions aim to promote such activity. Many of these focus on sport or supervised physical exercise programmes, games equipment or adult-organized competitions (Stratton, 2000; Pangrazi et al, 2003; Stratton and Mullan, 2005; Verstraete et al, 2006; Ridgers et al, 2007). However, the sustainability of these programmes, and the likelihood that children will continue to exercise in these ways after school, is questionable. There are also limitations in a reliance on interventions involving sport or exercise programmes.

Although sport may promote activity for children who enjoy it, many children have little interest or aptitude in sport. Most importantly, sport is not play and should not be seen as a substitute for it. In addition, 'many primary school teachers have low levels of perceived competence in teaching physical education and fundamental movement skills', and children who like school least are the 'least likely to be influenced by school-based programs' (Booth and Okely, 2005). Children with poor coordination, or who are overweight, are less likely to be interested in sport. Such children may already be less active. Many vigorous 'active' games, moreover, require children to wait their turn; hence, the total level of activity may be less than children engaged in unstructured play. To compound the problem for the less physically adept children, some sporting games have elimination rules, where the least capable children get the least activity.

Sport is not play and should not be seen as a substitute for it

There can also be clear gender biases in sporting or active games in school grounds. In schools dominated by designated sporting spaces (e.g. ovals or courts), these are typically dominated by boys, particularly boys who are more physically competent (Dyment et al, 2009). Girls, generally, and the less physically competent boys are often excluded from participating in these activities (Barbour, 1999) and, to some extent, become cast as outsiders. Thus, in school grounds where there are few non-sporting spaces, girls can be 'systematically excluded by boys from much of the space' (Paechter and Clark, 2007, p320).

Fixed play equipment (jungle gyms, slides, swings) are also of limited effectiveness in promoting physical activity in children (Dowda, 2009). In some cases, fixed play equipment increases the level of sedentary activity, when children get bored easily with equipment that they cannot manipulate or 'tend to congregate on and under the equipment' (Dowda, 2009, p265).

Research in a South Carolina preschool showed that when children played on fixed equipment, only 13 per cent of the observational intervals were spent in moderate/vigorous physical activity (Brown et al, 2009). In contrast, portable equipment (e.g. balls and tricycles) is associated with children being more active (Bower et al, 2008; Hannon and Brown, 2008). It is likely that the ability to manipulate the portable play items allows children to create their own challenges and variety in play. A school ground intervention that creates endless variety of challenges for children is described in the following section.

School Grounds Greening and Forest Schools

One school ground intervention that has been shown to be effective in encouraging physical activity levels among children is known as school ground greening (Dyment and Bell, 2007; Dyment et al, 2009), a practice that is widespread in schools in Canada (and other countries): 'by their very design, green school grounds encourage children to get moving in ways that nurture all aspects of their health and development' (Bell and Dyment, 2006, p465).

Green school grounds nurture children's health and development

As Dyment and Bell (2007, p464) explain, 'greening' includes 'a range of changes [to] school grounds, including naturalization, habitat restoration, tree planting, food gardening and similar efforts to bring nature back to school' (see Figure 4.3). Greening typically entails the transformation of both the design and the 'culture' of school grounds (e.g. the rules that govern play, the social dynamics among students and the role of supervisors). School ground greening can also contribute to a 'softening' of the landscape and diversifying of children's play repertoires, creating safer, more inclusive and less competitive play environments where there is less aggression, more civility and more cooperation. Such environments contrast with more conventional school ground design of open spaces of grass or asphalt bounded by fences 'to contain and control students, facilitate supervision and promote competitive sports' (Bell and Dyment, 2006, p464).

Similar to school ground greening is the 'forest school' initiative, where children are given regular opportunities for learning and play in a forest setting as a normal part of their school week. The forest school initiative uses the outdoor environment to allow children 'to experience a carefully monitored element of risk and to become more familiar with the natural world' (Gill, 2007, p65). Forest schools began in Sweden in the 1950s, and then spread across Scandinavia and to the UK during the 1990s. As Gill (2007, p65) argues: 'their growing number is a sign that risk aversion can be challenged'.

Green school grounds and forest schools may not promote high levels of vigorous activity found in competitive rule-based games and sports. Yet, recent research suggests that moderate levels of physical activity provide benefits in reducing obesity (Frank and Niece, 2005), and children's play in green spaces provides moderate levels of activity in a diversity of behaviours – climbing, running, hiding, socializing, constructive and imaginary play. Green school

Figure 4.3 Planting rice in a small rice field set up at the corner of the school ground

Note: The Japanese believe that it is essential for children to learn how rice grows and experience how much care is necessary before they can consume their main diet. The rice field also forms part of biology, social and moral studies.

Source: Motohide Miyahara

grounds provide an abundance of malleable, natural loose materials, including logs and stumps, branches, leaves and pebbles, seed pods and cones, and grass stems, providing opportunities for imaginative, cooperative and construction play (Tranter and Malone, 2004). School ground greening has the advantage over the interventions described above in that children find green school grounds intrinsically motivating. In other words, green school grounds promote play among preschool and primary school children. Play provides the motivation for children's sustained physical activity.

Play provides the motivation for sustained physical activity

An Alternative Intervention: Loose Materials Plus Risk Reframing

Two factors limit the potential for greening programmes to have a significant impact upon children's school grounds. First, there is the cost (and the time) involved in greening a school ground. While minor changes to school grounds may be made cheaply and quickly, major changes to school grounds can only be achieved with access to considerable levels of funding support (Dyment, 2005). This means that schools in low socio-economic areas are much less likely to be able to implement large-scale greening projects, and even those that do are less likely to be committed to a long-term project. The second limitation is that money spent on creating an elaborate green school ground means little if there is not an accompanying philosophical commitment to allowing children to make use of the advantages provided, including the manipulation of loose materials.

An alternative low-cost intervention that recognizes one of the strengths of school ground greening interventions – the role of loose materials that can be manipulated – has been trialled in a Sydney primary school, and is now being tested in several primary schools (Bundy et al, 2008, 2009). The importance of loose materials or 'loose parts' in children's play has long been recognized (Nicholson, 1971); research on natural environments as play spaces for children has often noted the importance of loose parts (Fjortoft and Sageie, 2000), and they are also seen as important in playground design and school grounds (Barbour, 1999; Malone and Tranter, 2003).

Loose materials provide a constantly varying source of stimulation for children, in contrast to the fixed play equipment or sporting materials that are designed for a specific use. Loose materials have no single or even obvious play purpose, and children must use their imagination to turn them into play objects. Large loose materials also require cooperative play when children are unable to move them by themselves. In the intervention described below, a range of loose materials was introduced into a school ground, providing one important component of the school ground greening intervention (see Figure 4.4).

Loose materials provide a constantly varying source of stimulation

However, just as green school grounds need to be accompanied by a school policy that allows children to make use of the grounds, so, too, loose materials by themselves are only of value to children for play if they are allowed (indeed, encouraged) to use them. Hence, it is important to overcome adults' perceptions that play equates to risk, and that risk should be minimized (especially in the school grounds): 'teachers and schools are also vulnerable to the expectation from some parents that education should be entirely free of risk' (Gill, 2007,

Figure 4.4 Play materials introduced into a Sydney, Australia, school ground to encourage social creative play

Source: Lina Engelen

p66). Thus, the intervention in this study employed a risk reframing exercise to address the phenomenon of surplus safety in the school grounds.

The intervention increased activity levels as well as social and creative play in the school playground by providing unstructured construction materials (e.g. hay bales, car tyres and plastic crates) for children to use as they wished with minimal interference from adults (Bundy et al, 2009). Data on activity levels from accelerometers, as well as data from the 'Test of playfulness' (Skard and Bundy, 2008), and information from interviews with teachers all indicated that children became more active, social and creative after the introduction of loose materials into the school ground.

The loose parts intervention also changed the social hierarchy in the school ground in a similar way to that reported in the study by Barbour (1999) above. After the introduction of loose materials, the dominant children were not always the strongest, most physically capable children who were best at sport or climbing. A new group of leaders emerged – the children with creativity and imagination – who could invent new ways of using the materials introduced into the school grounds. Another interesting observation was that the older children and younger children sometimes played together, often with older children making an elaborate play space out of loose materials. In one case, when older children had made a tunnel, other children delighted in crawling through the tunnel, and the older ones sometimes assisted younger children in getting through. Researchers' observations also indicate that gender differences in children's play were not as marked as in school grounds dominated by sporting activities.

Children's Views on What Makes a Good School

When primary school children in Canberra were asked to identify the things that make a good school ground, the themes that emerged as important to them were things that gave them more challenge in their play – both physical challenges as in climbing equipment or trees, and mental challenges, such as a maze or cultural activities and drawing (see Table 4.1). There was a strong desire among children to be allowed to climb, particularly to climb trees. Increased contact with nature was indicated in a desire for better grass and more trees in the school ground. There was also an identification of the need for more bubblers (drinking fountains), so that children can easily get a drink while they are playing – without having to disrupt their play with a long walk/ run back to the bubblers. Figure 4.5 shows a drawing by a year 4 child of a school ground as he would like it to be. Significantly, this drawing features trees (forest) and gardens, as well as a cubby (Australian term for den or fort) constructed by the children themselves. The drawing also features a swimming pool, not surprising given Canberra's hot dry summers.

Primary school children at Forbury School in Dunedin were asked about their views on what makes a great classroom (see Table 4.2). The children identified themes about the 'space' of the classroom – the light, the colour and the space. They particularly valued open space where creativity could occur, as well as quiet space for reading, and spaces of different sizes. They also had

Table 4.1 Children's views on what makes a great school ground

More interesting playground and more team sports.	School should have more shaded areas so when it rains we're not stuck inside.
Paint ball and zone alleys.	Let children climb trees.
Indoor swimming pool.	Swimming pool.
Flat handball and tennis courts.	More bubblers; trees we can climb.
Posts for oval; a maze.	More bubblers and more challenging equipment.
More bubblers (drinking water fountains) and trees.	Permission to climb trees; area to play chess or likewise.
Some senior equipment to give us a challenge.	Culture activities, cover over playground, pets.
More bubblers.	More bubblers, better playground and maybe better grass.
'Out of bounds' signs.	Covered oval and courtyard.
Areas for older kids, challenging equipment, more trees you're allowed to climb, more drawn-on pictures on the ground (e.g. handball areas).	More climbing bars that we're allowed to use and more sections just for seniors.

Figure 4.5 Drawing of Orana School ground, Canberra, Australia, as a child would like it to be

Source: nine-year-old child, Orana School, Canberra

Table 4.2 Children's views on what makes a great classroom

The classroom is bright and the teacher aid and teacher are nice.	Big and fun classroom.
When you need some time alone you can go anywhere.	A great class is about the massive work space.
Space in the classrooms; tiny space is the tent no one really goes there, but [...] .	The back because me and [other students] can do a map.
Games, great teachers and never having nothing to do.	A good classroom has a good learning environment, happy colours, open spaces and lots of learning tools available.
Our class is cool because we have the big office spaces.	The library because I like to read.
Library area because that's where you can go and read a book silently.	The great activities.
The mat is a great space; there's lots of room to sit and there are couches.	The art room makes a great classroom.
The classroom is great because of the art work.	A good classroom is a classroom that has space to play and space to hang things on the walls and space to put desks and more.
Big learning area.	A great classroom is warm and a big area
Areas that have our beautiful art where everyone can see it.	Fun activities room and a bigger desk for toys, and desk combinations to stop people stealing from desks.
What makes a classroom great is colour.	A good class is when you get along with the teachers and kids.
I want my classroom fun with easy worksheets.	Because our classroom has two rooms: one is the music room and the other is our normal room.
Tables and a whiteboard and chairs.	Size and shape: if a classroom was round the desks or tables would not fit; it needs to be a wee bit big and not round.

views on the social interaction in the classroom; in a great classroom they would have fun and 'get along with' their classmates and teachers.

Schools as Community Assets

Schools are an asset for the whole community

Schools and school grounds have undergone a variety of transformations over the last few decades. In some cases, children have played an active part in planning and making these changes. However, in most cases, children at school have been constructed as 'adults in training', and their interests as children in the 'here and now' have arguably not been well served, even by well-intentioned policies to keep them safe in their school grounds. Several strategies to make school grounds more child-friendly places have been implemented in cities around the world. The most valuable of these changes are those that allow children some control over their play space, either through contact with nature (school ground greening and forest schools) or through the freedom to play with a range of introduced loose materials. Most of the changes to children's schools have not challenged the fundamental concept of a school. Yet, the model of schooling has been critiqued, particularly for its adequacy in facilitating the environmental learning discussed in this chapter.

Steen (2003) provides a well-argued critique of the school as a model for education, arguing that there have been plenty of 'first-order changes', with very few 'second-order changes'. First-order changes 'improve the efficiency and effectiveness of what is currently done, without ... substantially altering the way that children and adults perform their roles', while second-order changes alter 'the fundamental ways in which organizations are put together' (Fullan, 1991, cited by Steen, 2003, p191). Steen argues that if schools are to be effective in educating children about the environment, and addressing ecological challenges, second-order changes are required. He identifies several limitations of the current models of schools, which he argues fit a mechanistic worldview, where phenomena are viewed in isolation, and technical fixes are identified in linear problem/solution terms (Steen, 2003). He explains that when children are centrally located in a school building that exists for the purpose of educating them, this limits their opportunities to gain direct experience of the world beyond the school, and teaches them that 'real learning happens inside and is composed of something other than their own natural observation' (Steen, 2003, p195). Because schools are segregated from the rest of society, children are separated from people of other ages. Schools operate for only four to six hours per day, implying that the other 18 to 20 hours are time when learning does not occur. The school is further broken down into single rooms, where children are restricted in their ability to learn from children outside their same-age peer group (Steen, 2003).

One way to challenge the current model of schools is to see schools as an asset (or service space) for the whole community, an issue that will be further explored in Chapter 7. We also return to schools in Chapter 10, looking at mobility (including the journey to school), and Chapter 13, where we discuss the role of schools in meeting a range of likely future challenges for our society. In the next chapter, we focus on the local neighbourhood as a space for children.

References

Adams, J. G. U. (1995) *Risk*, University College London Press, London

Barbour, A. C. (1999) 'The impact of playground design on the play behaviors of children with differing levels of physical competence', *Early Childhood Research Quarterly*, vol 14, no 1, pp75–98

Bell, A. C. and Dyment, J. E. (2006) *Grounds for Action: Promoting Physical Activity through School Ground Greening in Canada*, Toyota Evergreen Learning Grounds Program, www.evergreen.ca/en/lg/pdf/PHACreport.pdf

Bergen, D. (2002) 'The role of pretend play in children's cognitive development', *Early Childhood Research & Practice: An Internet Journal on the Development, Care and Education of Young Children*, vol 4, no 1, available at www.eric.ed.gov/PDFS/ED464763.pdf, accessed 12 January 2011

Booth, M. and Okely, A. (2005) 'Promoting physical activity among children and adolescents: The strengths and limitations of school-based approaches', *Health Promotion Journal of Australia*, vol 16, no 1, pp52–54

Bower, J. K., Hales, D. P., Tate, D. F., Rubin, D. A., Benjamin, S. E. and Ward, D. S. (2008) 'The childcare environment and children's physical activity', *American Journal of Preventative Medicine*, vol 34, no 1, pp23–29

Brown, W. H., Pfeiffer, K. A., McIver, K. L., Dowda, M. and Addy, C. (2009) 'Social and environmental factors associated with pre-schoolers' non-sedentary physical activity', *Child Development*, vol 80, no 1, pp45–58

Buchanan, C. (1999) 'Building better playgrounds: A project for parents?', *UAB Magazine (University of Alabama)*, vol 19, no 3, http://main.uab.edu/show.asp?durki=25353, accessed 12 April 2006

Bundy, A., Tranter, P. J., Luckett, T., Naughton, G., Wyver, S., Spies, G. and Ragen, J. A. (2008) 'Playful interaction: Occupational therapy for "all" children on the playground', *American Journal of Occupational Therapy*, vol 62, no 5, pp522–527

Bundy, A., Luckett, T., Tranter, P., Naughton, G., Wyver, S., Ragen, J. and Spies, G. (2009) 'The risk is that there is "no risk": A simple, innovative intervention to increase children's activity levels', *International Journal of Early Years Education*, vol 17, no 1, pp33–45

Burdette, H. L. and Whitaker, R. C. (2005) 'Resurrecting free play in young children: Looking beyond fitness and fatness to attention, affiliation, and affect', *Archives of Pediatric & Adolescent Medicine*, vol 159, no 1, pp46–50

Disinger, J. F. (1990) 'Needs and mechanisms for environmental learning in schools', *Educational Horizons*, vol 69, no 1, pp29–36

Dowda, M. (2009) 'Policies and characteristics of the preschool environment and physical activity of young children', *Pediatrics: Official Journal of the American Academy of Pediatrics*, vol 123, ppe261–e266

Dyment, J. E. (2005) '"There's only so much money hot dog sales can bring in": The intersection of green school grounds and socio-economic status', *Children's Geographies*, vol 3, no 3, pp307–323

Dyment, J. E. and Bell, A. C. (2007) 'Active by design: Promoting physical activity through school ground greening', *Children's Geographies*, vol 5, no 4, pp463–477

Dyment, J., Bell, A. and Lucas, A. (2009) 'The relationship between school ground design and intensity of physical activity', *Children's Geographies*, vol 7, no 3, pp261–276

Elliot, S. (2008) 'Risk and challenge: Essential elements in outdoor playspaces', *Every Child*, vol 14, no 2, pp12–13

Evans, J. (1995) 'Children's attitudes to recess and the changes taking place in Australian primary schools', *Research in Education*, vol 56, pp49–61

Evans, J. (1997) 'Rethinking recess: Signs of change in Australian primary schools', *Education Research and Perspectives*, vol 24, no 1, pp14–27

Evans, J. (1998) 'School closures, amalgamations and children's play: Bigger may not be better', *Children Australia*, vol 23, no 1, pp12–18

Evans, J. and Pellegrini, A. (1997) 'Surplus energy theory: An enduring but inadequate justification for school breaktime', *Educational Review*, vol 49, no 3, pp229–236

Fjortoft, I. and Sageie, J. (2000) 'The natural environment as a playground for children: Landscape description and analyses of a natural landscape', *Landscape and Urban Planning*, vol 48, no 1–2, pp83–97

Frank, L. and Niece, J. (2005) *Obesity Relationships with Community Design: A Review of the Current Evidence Base*, Heart and Stroke Foundation, Ottawa, Canada

Frost, J. L. (2007) 'An historical imperative to save play', in E. Goodenough (ed) *Where Do the Children Play?*, Michigan Television, Michigan

Gill, T. (2007) *No Fear: Growing Up in a Risk Averse Society*, Calouste Gulbenkian Foundation, London

Hannon, J. C. and Brown, B. B. (2008) 'Increasing pre-schoolers' physical activity intensities: An activity-friendly preschool playground intervention', *Preventive Medicine*, vol 46, no 6, pp532–536

Hart, R. (2002) 'Containing children: Some lessons on planning play from New York City', *Environment and Urbanization*, vol 14, no 2, pp135–148

Herrington, S. and Lesmeister, C. (2006) 'The design of landscapes at child-care centres: Seven Cs', *Landscape Research*, vol 31, no 1, pp63–82

Honoré, C. (2009) *Under Pressure: Putting the Child Back in Childhood*, Vintage Canada, Toronto

Isenberg, J. P. (2002) 'Play: Essential for all children', *Childhood Education*, vol 79, pp3–39

Johnson, J. M. (2000) *Design for Learning: Values, Qualities and Processes of Enriching School Landscapes*, American Society of Landscape Architects, www.asla.org/latis1/LATIS-cover.htm, accessed 20 November 2001

Kytta, M. (2004) 'The extent of children's independent mobility and the number of actualized affordances as criteria for child-friendly environments', *Journal of Environmental Psychology*, vol 24, no 2, pp179–198

Lambert, E. B. (1999) 'Do school playgrounds trigger playground bullying?', *Canadian Children*, vol 24, 1, pp25–31

Malone, K. and Tranter, P. J. (2003) 'Schoolgrounds as sites for learning: making the most of environmental opportunities', *Environmental Education Research*, vol 9, no 4, pp283–303

Mayes, T. and Chittenden, M. (2001) 'Play areas to teach lost art of risk-taking', *The Sunday Times*, 24 June, p24

Moore, R. C. (1986) *Childhood's Domain: Play and Place in Child Development*, Croom Helm, London

Moore, R. and Wong, H. (1997) *Natural Learning: The Life History of an Environmental Schoolyard – Creating Environments for Rediscovering Nature's Way of Teaching*, MIG Communications, Berkeley, CA

Nicholson, S. (1971) 'How not to cheat children: The theory of loose parts', *Landscape Architecture*, vol 62, no 1, pp30–35

Paechter, C. and Clark, S. (2007) 'Learning gender in primary school playgrounds: Findings from the Tomboy Identities Study', *Pedagogy, Culture and Society*, vol 15, no 3, pp317–331

Pangrazi, R. P., Beighle, A., Vehige, T. and Vack, C. (2003) 'Impact of promoting lifestyle activity for youth (PLAY) on children's physical activity', *Journal of School Health*, vol 73, no 8, pp317–321

Play England (2008) *Managing Risk in Play Provision: 4 Position Statement, National Children's Bureau*, www.playengland.org.uk/media/120462/managing-risk-play-safety-forum.pdf, accessed 11 January 2011

Ridgers, N. D., Stratton, G., Fairclough, S. J. and Twisk, J. W. R. (2007) 'Children's physical activity levels during school recess: A quasi-experimental intervention study', *International Journal of Behavioral Nutrition and Physical Activity*, vol 4, no 19, doi:10.1186/1479-5868-1184-1119

Singer, D. G., Golinkoff, R. M. and Hirsh-Pasek, K. (2006) *Play = Learning: How Play Motivates and Enhances Children's Cognitive and Social–Emotional Growth*, Oxford University Press, Oxford

Skard, G. and Bundy, A. C. (2008) 'Test of playfulness', in L. D. Parham and L. S. Fazio (eds) *Play in Occupational Therapy for Children*, Mosby, St Louis

Smith, G. (1992) *Education and the Environment: Learning to Live with Limits*, State University of New York Press, Albany, NY

Steen, S. (2003) 'Bastions of mechanism, castles built on sand: A critique of schooling from an ecological perspective', *Canadian Journal of Environmental Education*, vol 8, spring, pp191–203

Steinsvik, R. M. (2004) 'Playgrounds in kindergartens, schools and residential areas', in *Challenging Winter Frontiers*, Anchorage, Alaska, www.upea.com/winter/playgrounds.pdf, accessed 12 April 2006

Stephenson, A. (2003) 'Physical risk-taking: Dangerous or endangered', *Early Years: An International Journal of Research and Development*, vol 23, no 1, pp35–43

Stratton, G. (2000) 'Promoting children's physical activity in primary school: An intervention study using playground markings', *Ergonomics*, vol 43, no 10, pp1538–1546

Stratton, G. and Mullan, E. (2005) 'The effect of multicolor playground markings on children's physical activity level during recess', *Preventive Medicine*, vol 41, no 5–6, pp828–833

Thomson, S. (2002) 'Harmless fun can kill someone', *Entertainment Law*, vol 1, no 1, pp95–103

Thomson, S. (2003) 'A well-equipped hamster cage: The rationalisation of primary school playtime', *Education 3–13*, vol 31, no 2, pp54–59

Thomson, S. (2007) 'Do's and don'ts: Children's experiences of the primary school playground', *Environmental Education Research*, vol 13, no 4, pp487–500

Timmons, B. W., Naylor, P. and Pfeiffer, K. A. (2007) 'Physical activity for preschool children – how much and how?', *Applied Physiology, Nutrition and Metabolism*, vol 32, pp122–134

Titman, W. (1992) *Play, Playtime and Playgrounds: Key Issues for Teachers, Supervisors and Governors of Primary Schools*, Learning through Landscapes, World Wide Fund for Nature, UK

Titman, W. (1994) *Special Places, Special People: The Hidden Curriculum of School Grounds*, Learning through Landscapes/WWF, Cambridge, UK

Tranter, P. J. (2006) 'Overcoming social traps: A key to creating child friendly cities', in B. Gleeson and N. Sipe (eds) *Creating Child Friendly Cities: Reinstating Kids in the City*, Routledge, New York, NY, pp121–135

Tranter, P. J. and Malone, K. (2004) 'Geographies of environmental learning: An exploration of children's use of school grounds', *Children's Geographies*, vol 2, no 1, pp131–155

Tranter, P. and Malone, K. (2008) 'Out of bounds: Insights from Australian children to support sustainable cities', *Encounter: Education for Meaning and Social Justice*, vol 21, no 4, pp20–26

Uzzell, D. (1990) 'An environmental psychological perspective on learning through landscapes', Appendix 5 in Adams, E. (ed) *Learning Through Landscapes: A Report on the Use, Design, Management and Development of School Grounds*, Learning Through Landscape Trust, Winchester, UK

Verstraete, S. J. M., Cardon, G. M., De Clercq, D. L. R. and De Bourdeaudhuij, I. M. M. (2006) 'Increasing children's physical activity levels during recess periods in elementary schools: The effects of providing game equipment', *European Journal of Public Health*, vol 16, no 4, pp415–419

White, R. and Stoecklin, V. (1998) *Children's Outdoor Play and Learning Environments: Returning to Nature*, White Hutchinson Leisure & Learning Group, www.whitehutchinson.com/children/articles/outdoor.shtml, accessed 3 December 2001

Wilde, G. J. S. (1998) 'The concept of target risk and its implications for accident prevention strategies', in A. M. Feyer and A. Williamson (eds) *Occupational Injury: Risk, Prevention and Intervention*, Taylor and Francis, London, pp82–105

5
Neighbourhood

> In a natural city this is what happens. Play takes place in a thousand places – it fills the interstices of adult life. As they play children become full of their surroundings. How can a child become filled with his surroundings in a fenced enclosure? He can't. (Alexander, 1966, p12)

What Is a Neighbourhood?

Neighbourhood at its simplest is the physical space or locality in which people live: it provides the houses, shops, roads, parks, services, schools and other facilities upon which its residents depend. 'Community', which is often used interchangeably with 'neighbourhood', is a social group connected in some way, but who may or may not live in the same geographic location. Traditionally, community and neighbourhood have been seen as closely related; but some question whether this link is eroding as people's lives are increasingly scattered across broad expanses of the city (Kearns and Parkinson, 2001) and ask whether neighbourhood still matters. Others argue that neighbourhood remains an integral and valued part of urban life, where 'relationships between neighbours continue to confound predictions of their redundancy' (Crow et al, 2002, p141). For children, neighbourhood plays an essential role in their lives and its demise has serious negative impacts upon their life worlds. Neighbourhoods are places where they begin to encounter the world outside the home, where they make their first independent forays and where they become part of wider public life. As Bartlett et al (1999, p123) explain, this move to independence can only happen if the neighbourhood base is itself a place that provides good experiences:

Neighbourhoods are places where children become part of wider public life

> Ideally a neighbourhood should be a place where children can play safely, run errands, walk to school, socialize with friends and observe and learn from the activities of others. When neighbourhoods provide a secure and welcoming transition to the larger world, children can gradually test and develop their competence before confronting the full complexity of the city ... local neighbourhoods can be complex environments, calling for more sophisticated skills, understandings and choices than are necessary within the household ... also provide the opportunity for children to begin to understand, accept and ideally enjoy differences, a critical part of their development as tolerant, responsible citizens.

Neighbourhoods differ noticeably in their geographic form, their social and economic composition, their built fabric and their history. Neighbourhoods also diverge markedly with regard to whether they work for or against children's well-being. This chapter examines children's neighbourhood relationships, the factors that make good neighbourhoods for children, examining both traditional neighbourhoods and more recent neighbourhood development initiatives.

Child-Friendly Neighbourhoods

What makes a neighbourhood good for children is little different from what makes a neighbourhood good for the general population, and comprises two main elements: a good social environment and a good physical environment. It provides a sense of well-being, belonging and social connection, and is a place where people feel safe, valued and supported. It exudes 'neighbourliness', a concept that – while familiar – is seldom directly defined. 'Neighbourliness' refers to daily interchange between people; it involves socializing, giving assistance. It is generative in that neighbourly places encourage neighbourliness, where people connect more with the people with whom and the place in which they live (Pilch, 2006). Children can act as a catalyst for this socialization (see Figure 5.1), as Gill (Gill, 2007, p7) notes: 'the presence of children playing in the street can be seen as the litmus test of the level of community cohesiveness in a neighbourhood'. It is recognized that where children interact more in the neighbourhood, so too do the adults (Offer, 2007, p1125). A key concept in this regard is that of social capital, defined in simple terms as the positive social resources that are 'the product of social relationships' (Bassani, 2007 p17), and implying trust, reciprocity, shared values and the existence of strong social networks. Coleman (1988), who many regard as the architect of the theory of social capital, recognized the role of children in this regard and believed that communities where parents and children exhibit strong connections also develop good social capital.

An increasing problem many families face is that of social isolation, where parents do not know their neighbours and relatives may be far away. This

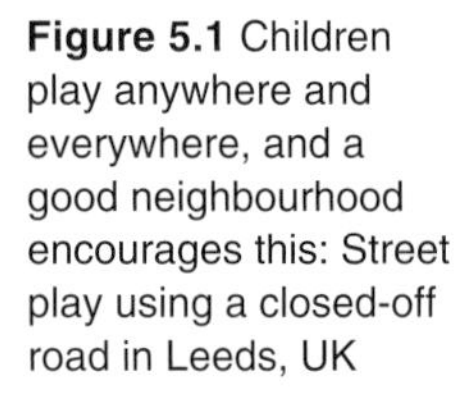

Figure 5.1 Children play anywhere and everywhere, and a good neighbourhood encourages this: Street play using a closed-off road in Leeds, UK

Source: Claire Freeman

increases pressures on the family who has to take full responsibility for all aspects of childcare. Families need to be able to rely on neighbours for help, not just to borrow the proverbial cup of sugar or to feed the cat, but also to act as back-up when child care arrangements fail, to provide places for children and families to have time out from each other and to provide alternative role models, and to broaden children's understandings and experience of society. Furedi (2008, p38) sees the isolation of parents as fuelling what he calls 'paranoid parenting', where danger is seen as ever present and parents are anxious and likely to overreact. Parental isolation is not inevitable and can be overcome. Good design, for example, where homes face each other, have low fences, face directly onto and relate to the street, and where houses are connected to communal gathering places, encourages meetings between people. In the existing urban form, retrofitting, through safe street initiatives, encouraging mixed-use local shopping centres and improved public transport, can act as positive encouragements in this respect too. Communities can also address isolation through community development initiatives in churches, temples, schools, health services, sports organizations, clubs, community and special interest groups, events and celebrations (see Figure 5.2). The emphasis

Figure 5.2 Celebrations and festivals are an important part of public life and bring communities together: Dragon parade in Japan

Source: Motohide Miyahara

Table 5.1 Alberta Premier's Council, Canada: Community self-assessment checklist

Family-friendly communities	Child-friendly communities
Neighbours welcome families as community members.	Children have access to all members of their extended family.
People know their neighbours.	Community members value and care for children.
Appropriate community events and celebrations include all family members.	Children contribute and are an active part of community life.
Community events are sensitive to, and reflective of, the diversity of family types and multicultural aspects of the area.	Children are present and participate in activities and events organized by the community.
Families know about community resources and activities.	Programmes are available to support the growth and development of young children and to support parents.
Locations are established for families to recycle toys, clothing, equipment, etc.	Children in trouble know where to go to for help.
Families have access to a community meeting house or venue.	Local restaurants, shopping malls and businesses have change rooms, play areas and staff happy to serve children.
Food stores, public library, swimming pool, etc. are within easy reach of the neighbourhood.	Street crossings are safe for children and walkways are clean and well lit.
Houses and neighbourhoods are designed to meet family needs.	Play areas are safe and visible to parents.
Neighbours support each other.	Quality childcare alternatives are available and accessible.
Public transportation systems and community-based systems of support are accessible and available to all members of the community.	Facilities are accessible to strollers, carriages, wheelchairs, etc.
Neighbours are available and willing to help in an emergency.	Safe places are available for children to participate in unstructured activities.
Mediation is available to settle disputes between neighbours.	Supervised arts and crafts programmes are available to encourage children's creativity.
There are natural gathering places for people of all ages.	Activities are available for children without concern for cost.
There is a sense of pride and cooperation in the community.	

on neighbourhoods as social entities and as social supports is reflected in the checklist designed for Alberta, Canada, in assessing whether communities are family and child friendly (see Table 5.1). This checklist prioritizes community life, social interchange and support. Another useful methodology is the *How Friendly Is My Community* workshop pack produced by UNESCO Growing up in Cities–Asia Pacific (2007), which can be used by agencies wanting to work with children at the local, community and neighbourhood level.

The built form also affects children's health. A good neighbourhood which provides good social interactions, and which values and respects children, will enable them to develop good mental health. Similarly, good neighbourhoods can also support physical health through providing accessible neighbourhoods that encourage walking, cycling, adventure and exploration; that are clean and unpolluted; that are safe from traffic, violence and other harms; and that provide accessible usable open space and access to services such as childcare, school, doctors and public transport. Unfortunately, many neighbourhoods are socially and physically inadequate in a number of ways:

- *physically poor* – inadequate housing promoting ill health; buildings and services in a state of disrepair and decline;
- *unsafe* – from violence, traffic, pollution;

- *socially challenging* – lacking in neighbourliness, discouraging contacts and social cohesion, including people unsympathetic to children;
- *unstable* – neighbourhoods with fluid populations, insecure tenure, refugee camps, trailer parks, neighbourhoods under threat, proposed for demolition, road building;
- *powerless* – neighbourhoods where children feel ignored and unsupported, with few rights and unable to participate in decision-making;
- *isolated* – neighbourhoods where children are isolated from each other and/or from the wider community;
- *divided* – elements of the community do not value or respect other elements of the community; residential areas may be segregated with high levels of conflict and alienation;
- *bland* – lack vibrancy; socially and physically unstimulating; limited activities.

Good neighbourhoods embrace difference; they value, for example, the child's family, cultural, racial and religious affiliations. Social well-being is closely linked to the design of the built neighbourhood, which can encourage or restrict opportunities to develop social connectedness. Where children live in deprived neighbourhoods, this can severely affect their 'life chances' through poverty, lack of educational and other opportunities, and exposure to strife and violence. Children living in neighbourhoods not usually considered to be deprived can also have negative experiences. In particular, children living in higher-income neighbourhoods, in larger homes behind walls and isolated from neighbours and the community, can experience extreme isolation and associated poor mental health (see Figure 5.3). Children's development and emotional health are

Figure 5.3 Homes like this one in Dunedin, New Zealand, contribute to social isolation and are especially forbidding to children, with their high walls and 'phone' entry

Source: Claire Freeman

also inhibited in neighbourhoods that have become so 'sanitized' that they offer little in the way of exploration, contact with nature, adventure and challenge.

Children in challenging neighbourhoods develop substantial but largely unrecognized skills

Children in what are commonly seen as challenging neighbourhoods can also develop substantial but largely unrecognized skills. These skills are clearly revealed in the UNESCO Growing Up in Cities programme, where diverse neighbourhoods that would be classed as disadvantaged include Canaansland squatter camp in South Africa; Sathyangar informal settlement, Bangalore, India; Fruitvale, a deprived, predominantly ethnic neighbourhood in Oakland, US; Braybrook, also deprived, with a high ethnic population in Melbourne, Australia; and post-communist Powisle, Warsaw, in Poland (Chawla, 2002). While the children in these neighbourhoods do suffer from poor physical conditions and have limited opportunities to go outside their immediate neighbourhood, they often confound the negative preconceptions associated with such neighbourhoods. These children frequently show immense eagerness in taking advantages of opportunities as they arise. One example is the overwhelming enthusiasm for participatory involvement shown by children in Canaansland. Research organizers had planned to meet on a Saturday with a small group of 10- to 14-year-old children to look at evaluating the neighbourhood. On the day, organizers were confronted with a mass influx of children of all ages arriving at the venue and wanting to participate. The children who did participate were keen to get on with the work and proposed some of the data collection methods. Despite the obvious problems of the neighbourhood, most children 'spoke of their shacks in Canaansland as safe and happy places' (Swart-Kruger, 2002, p119). The stories provided by the children from these 'challenging' neighbourhoods reveal their resilience, deep local knowledge, independence, advanced life skills, and often highly advanced social skills.

This enthusiasm was also experienced by the author when undertaking research in schools in Fiji, where every child in the classes approached to be part of a research project brought back a completed parental consent form the next day to ensure that they could participate. Other examples of resilience and acquisition of skills in the UNESCO Growing up in Cities programme were noted where some of the children from homes where the adults lack English language skills act as mediators with the external world for the 'immigrant' adults in their lives. Overall, the children show great insight into their lives and neighbourhoods, and are well aware of their neighbourhood's shortcomings, as indicated in the following extract from Meg talking about her Braybrook neighbourhood in Melbourne: 'When I'm walking around, I don't go very far and not near the old abandoned houses ... I know where to avoid, you've got to. Even which houses, like the druggies' houses or where they've got big dogs' (Malone and Hasluck, 2002, p97). As Meg reveals, children are adept at working within and around neighbourhood constraints in ways that, while not diminishing the physical and social problems present, do build on their own innate social, cultural and communal skills. For children living in these challenging neighbourhoods, the problems are generally known, allowing children to develop avoidance strategies. Where children lack neighbourhood knowledge – as is the case for increasing numbers of children in better-off, 'safe' Western neighbourhoods – their lack of knowledge means that the world outside their home can be for them a far more intimidating and uncomfortable place. Different countries over time have adopted different ways of designing neighbourhoods that meet the needs of children and their families with quite differing degrees of success. Housing provision and design informs the next section.

Building Neighbourhoods: The Rise of 'Family Housing'

A neighbourhood is a collection of homes. Most children live in homes whose design has been determined by some external agency, usually in response to political and economic decisions and subject to architectural, planning and building norms and standards. Rarely is the house in which a child lives built to their own specification or reflective of their own family needs. Mass-produced housing emerged during the Industrial Revolution, with large-scale housing built to predetermined designs and produced on a scale never before seen. With population movement to the cities, demand for new housing increased. Architects, planners and builders responded with house and neighbourhood designs that have become enshrined in popular architectural and planning folklore for both good and ill. Ebenezer Howard's garden city model is perhaps the best known, with its quintessential low-density solid brick homes, surrounded by gardens and tree-lined avenues. Another innovative housing estate design was the Radburn layout of 1929 (by Clarence Stein and Henry Wright), later exported across the world and often used in modern private housing developments. In this design houses were grouped around open space, with vehicle traffic directed away from the front and to the rear of houses. Key living and sleeping rooms faced the open space. For children, both the garden city and Radburn models combined plenty of open space with easily accessible community living. The reverse was true for some of the more infamous (usually public) high-rise, high-density housing designs that followed. As housing markets, both social and private, developed during the 20th century, so, too, did the popularity of the 'family dream home'. Homeownership became a more realizable aspiration for families. In Australia and New Zealand, this 'dream' took the form of the quarter acre section; in the US, the suburban home; and in the UK, the suburban housing estate on the city fringe.

For many families on low incomes, a privately owned home was never going to become a reality, and the post-World War II era saw massive social (public) house building internationally – housing that ranged from excellent to abysmal in quality. In the US, perceptions of public housing tend to be dominated by some of the abject failures of mass public housing – developments such as the infamous Pruitt-Igoe Estate in St Louis. First occupied in 1954, it comprised nearly 3000 small apartments, housing mostly poor black families, and was based on a Le Corbusier-type high-rise, high-density design with no personal outdoor space for families and their children. Never fully occupied, it was demolished in 1974 after it was deemed unfit for habitation.

Massive public house building internationally ranged from excellent to abysmal

Between 1965 and 1974, Sweden had its *Miljonprogrammet* programme to build 1 million new houses (widely varying in height and density). These tended to be on new estates linked to the existing cities by mass public transport (Hall and Vidén, 2005). Rejected by many of the families destined for this housing, these housing estates have been characterized by poverty, crime and alienation, providing limited life opportunities for children and their families.

Some of the poorer social housing developments have left an enduring legacy for the children unfortunate enough to be condemned to growing up in some of them, especially the more soulless, cramped, spatially deficient high rises that proliferated during the 1960s and 1970s.

More enduring in its popularity has been low-density suburban housing, both social and private, built with the aim of providing spacious, healthy housing for 'happy families'. The premise was that healthy homes with healthy children would produce happy, contented families and future citizens. Suburban housing is not without its own problems though. Young and Wilmot (1957), authors of the classic study *Family and Kinship in East London*, lived among families in Bethnal Green, London, whose homes were destined for slum clearance and rehousing in the new suburban housing estates in Essex. On moving to the new estates, the sense of community contact with extended family and neighbourliness characteristic of Bethnal Green was lost. While children gained space, they lost much of their sense of belonging, and their care became the responsibility of the nuclear rather than the extended family.

As families shifted to the suburbs children lost access to the vitality and services of the city

As families shifted to the suburbs in the UK, Australasia, Canada and the US, so children have lost access to the vitality and services of the city (see Chapter 6). At the same time as some families moved out, their inner-city homes have become occupied by other newer immigrant families in an ongoing internationally observed cycle of housing occupation and reoccupation. Some immigrant communities inhabit clearly identifiable neighbourhoods close to the city: the Vietnamese in West Melbourne, Australia; Bangladeshi families in the East End, London; Japanese communities in Hong Kong; Filipino communities in Ontario, Canada; Turkish communities in the Kreuzberg area of Berlin; and the many Chinese communities in cities around the world. The experience of children from newly immigrant communities can be starkly different from those of more established communities – physically more deprived, perhaps, but socially often rich. Children living in poorly designed higher-density rental accommodation not intended as family housing is a growing problem in cities as diverse as Auckland, New Zealand, and London, UK. Apartments and medium-density housing need to be designed with children in mind – not, as is frequently the case, as cheap temporary housing for families with no other housing options.

Designing the Family Neighbourhood

Housing provision and design have experienced a number of developmental phases, the key ones summarized in Table 5.2. While these phases are indicated sequentially, they are not mutually exclusive in that different housing types can be built simultaneously; while being commonly associated with particular time periods, house types are not necessarily confined to these periods. In different countries, housing development has taken different pathways; but what is common is the tendency for housing trends and designs to cross national and cultural borders as the international proliferation of tower block housing and the suburban house style indicate. Some housing styles have proven to be more robust and appreciated than others. The single- or two-storey house with its own private garden, whether a stand-alone house or attached to others, as in terrace or town house developments, remains universally popular. Medium density seems to work better for children and their families than either low

Table 5.2 Phases in European house development

Approximate timing		Housing character
Late 1880s to 1930s	Industrial Revolution mass-built workers' housing	Small workers' homes, terraced housing, small connected workers' cottages, tenement housing
1940s to1960s	Pre- and post-World War II rebuilding	Early suburban development Social housing estates – mostly greenfield Major apartment building in Europe – large uniform blocks in Eastern Europe and parts of Asia
1960s to 1970s	Slum clearance	Tower blocks Social housing, greenfield estate developments
1980s	Mass build private housing estates	Smallish individual, often stand-alone, duplex or attached townhouses Ongoing suburban development
1990s	Infill housing in cities	Small estates with mixed housing Housing on regenerated sites, such as waterfronts Medium- to high-density housing
2000 onwards	Outer-city medium-density planned estates	Higher-density housing with a trend towards single, couple and small family occupancy housing in countries with previously low density

density (often associated with sprawl and isolation) or high density (often associated with high rise, congestion and lack of open space). Housing density is reflective of location, densities being characteristically highest in city centres and lowest in more recently urbanized countries such as the US, Australia, Canada and New Zealand, and highest in Asian countries such as Japan, and European countries such as The Netherlands.

Is there an ideal house style for children? Some housing developments have proved better for children than others. Some of the better-built older terrace/ row housing continues to provide good housing for children in Europe. While the density of housing encourages neighbourly interaction, the lack of garden space, together with the increasing street traffic, which removes the street as a play space, can be problematic. Some terrace/row housing areas have been successfully revitalized through Home Zone areas, liveable streets, '*woonerf*' (in The Netherlands) and other traffic-calming initiatives (see Figure 10.6 in Chapter 10). Slum clearance provided immense opportunities for families to be provided with better-quality homes. Indeed, many of the new housing estates with their low- to medium-density housing, private gardens and traffic-calming designs, such as culs-de-sac, did provide good family homes (see Figure 11.2 in Chapter 11), even if some estates were distanced from city centres and services. However, for some families, slum clearance, with its removal to high-density housing and Eastern bloc housing developments, resulted in many children living in totally unsuitable high rise: monolithic, unappealing, often cramped flats with minimal access to outdoor play space. These designs were the nadir of housing provision for children, and unfortunately persist in many urban landscapes that still house some of the most deprived families. While the inadequacy of these poor-quality high-rise developments for children is universally acknowledged, less consensus has been reached on the legacy and appropriateness of a more common and enduring house style: the low-density suburban house. As with all housing types, quality is variable and there have been many examples of poor-quality suburban housing, hastily built, using poor-quality materials, with repetitive uninspiring design and providing few amenities, although high-quality suburban housing has also been constructed.

Since the 1990s and continuing into the 21st century, there have been significant new directions in family housing development. Factors that have led to the development of these newer housing forms include considerations around land price, encouraging building at higher densities, attempts to promote more compact cities and to reduce sprawl, environmental considerations, and the need to build more sustainable homes and a desire to recapture a sense of community in housing. The environmental/green city movement has been influential in bringing to the fore considerations around building homes that not only have fewer negative environmental impacts, but also encourage socially sustainable and healthy lifestyles (see Low et al, 2005). A number of development initiatives and trends have emerged. Most are positive, recognizing that housing works well at higher densities, in areas where there is a mix of facilities which are accessible by walking, have low motorized traffic densities, and where houses are designed to encourage interaction and neighbourliness (see Figure 5.4). A number of more recent housing developments are identified in Table 5.3, together with their suitability for children. While most have positive benefits, it is worrying that family housing is still being built that is prejudicial to the well-being of children and their families, as in some of the gated community and newer upmarket housing developments.

Figure 5.4 Good neighbourhoods that integrate built and natural features are aesthetically appealing and encourage walking and cycling, which works well for families (Osaka, Japan)

Source: Diane Campbell-Hunt

Table 5.3 New housing development types

Housing type	Location	Description	Child benefits
Eco-housing and eco-villages	Various but particularly strong in Northern Europe	Housing and community developments that emphasize environmental well-being and encourage more communal lifestyles	Emphasis on good open space provision, outdoor lifestyle and opportunities for social interaction with neighbours and other children
New Urbanism/smart growth	US, UK	Urban village-type developments; housing at higher densities	Higher-density housing encourages social interaction and can be socio-economically homogeneous; access is economically determined
Collective/shared housing	Various but particularly strong in Northern Europe	Unrelated residents share a single building or collection of buildings	Provides social interaction opportunities, can be more affordable for families, and can work especially well for children in non-nuclear families; access to communal amenities
Inner-city regeneration	Older industrial countries	Medium- to high-density housing encouraging repopulation of the inner city	Usually tends to be dominated by professionals and adults without children; houses aren't necessarily matched by provision of services and open space for children; some are excellent examples of vibrant social housing
Upmarket housing developments	Australia, New Zealand, US	Low to medium density; some estates have large houses taking up most of the land area; high surrounding fences/ walls; sometimes called 'McMansions'	Lack community ethos; limited play space around the house; walls and fences inhibit social interaction
Gated developments	Various but especially strong in Australia, US	Medium density; access restricted to residents	Socially and economically homogeneous; often not intended as family environments, so limited tolerance for children's play activities; difficult for children to access the wider society

Design is important and should reflect societal, cultural, geographic, historic and even economic imperatives. There is, however, no universal design for a good neighbourhood; a suburb doesn't necessarily provide a better neighbourhood for a child than a high-density inner-city one. The website for SickKids (the Canadian Hospital for Sick Children), affiliated with the University of Toronto, includes a list of questions that parents should ask when considering buying a house in a neighbourhood, indicating the recognition that there is a connection between children's physical and mental health and well-being and children's home environments (www.sickkids.ca/). What this list reveals is that good design is intimately connected with good health and caring, friendly and social neighbourhoods (see Box 5.1).

Box 5.1 SickKids Toronto checklist: Location, location, location – choosing a neighbourhood

If you're considering a move, ask your kids for their input. What features are important to them? What do they want in their neighbourhood? While some of them seem fairly obvious and some may not be priorities for you, the following questions can help you to determine whether a neighbourhood will be a healthy, friendly place for you and your children to live in.

Safety

- Is there a lot of traffic? What is the speed limit?
- Are there sidewalks on at least one side of every street?
- Are there bike lanes?
- Are there narrow streets to slow down drivers and help pedestrians and cyclists cross?
- On busier streets, are there many crosswalks and traffic lights?
- Are there 'eyes on the street' – neighbours and workers who will keep an eye out for trouble and be able to give help if needed?
- Do homes have front porches and windows facing the street?
- Is there adequate street lighting?

Scale

- Are there high, blank walls, or is the streetscape welcoming?
- Are there playgrounds, alleys and front porches?
- Is land used efficiently, with narrow roadways and reasonably sized lots?
- How far are the houses set back from the street?
- Are the blocks long or short? Longer blocks mean longer detours.
- Is it a pleasant place to walk or jog?

Tradition

- Are there monuments, landmarks or natural areas that can anchor kids to their community?
- What are the future development plans for the area?

Accessibility

- Is it close enough to where children need and want to go (schools, parks, playgrounds, recreational facilities, places of worship, stores, libraries, movie theatres, friends and family) for them to walk or bike there?
- For that matter, is it close enough for parents to walk, cycle or take public transit to work?
- Can you use your car less or not at all?
- Is it cut off by a major road or highway?
- Is it near public transportation that goes somewhere useful, or will kids have to take three different buses to get where they want?
- Are there bike paths that go somewhere?
- Are there places to park your bike when shopping or going to the library?
- Do other people walk or cycle?

Integration

- Do other kids live nearby?
- How easy is it for kids in the neighbourhood to play together in a casual, unstructured fashion?
- Can you and your family get to know neighbours and local shopkeepers?
- Does the neighbourhood have a mix of features such as schools, parks, recreational facilities, places of worship, stores, a library, doctors, dentists, and opportunities for after-school or summer jobs?
- Do people of different ages and backgrounds live in the area?
- Have natural areas in the neighbourhood been preserved?
- If your housing needs change (e.g. if you have another child or your children leave home), will you be able to stay in the neighbourhood?
- Is there a mix of different housing types available – big houses, small houses, apartment buildings?

Building Better Neighbourhoods

Given the importance of the physical form in determining the interactions between community residents, this chapter has devoted much attention to the built form, notably housing development and the relationship between houses. Good form encourages interaction between residents and promotes the visibility of children in society. The importance of socialization was identified in the UNESCO Growing Up in Cities project: 'Neighbourhoods, towns and cities should be places where children can socialise, observe, and learn about how society functions, and contribute to the social fabric of their community' (UNESCO, 2007, p5). But children cannot take responsibility for advancing communal relationships alone; they need to be part of a wider neighbourhood ethos that encourages and values sharing, caring and interface between its residents. This is why initiatives such as Growing Up in Cities are so important in working with children and their communities in participatory ways that transcend age, economic, religious and other community divides. Extensive research and trials on working with children in neighbourhoods (photo mapping, community walks, behaviour mapping, focus groups, and community observations; see Driskell, 2002, for a full account) reveal children's keen understanding of, reliance on and commitment to neighbourhood. Research by Freeman (2010) found that regardless of where they lived, and regardless of the varied reputations of neighbourhoods and how they looked to outsiders, children liked the homes and neighbourhoods in which they lived. Children are positive about neighbourhoods in a way that adults often are not. They are prepared to live within the constraints of the neighbourhood, but also challenge its limitations and find ways to make their lives work despite the constraints.

Children need to be part of a wider neighbourhood ethos that values sharing

Neighbourhood planning needs to nurture residents' well-being: to emphasize the human scale, to champion streets, not highways, corner shops, not shopping malls, local swimming pools and not merely leisure centres and water-worlds; to encourage mixed-use not single-use streets; and to have neighbourhood schools attended by neighbourhood children, rather than have parents driving children to distant, large amalgamated school complexes. It needs to champion neighbourhood green spaces, pocket parks, cafés, eateries and people-watching spaces that encourage lingering and 'spontaneous encounters'. The principles for good neighbourhoods are well known: Ebenezer Howard in 1902, writing his *Garden Cities of To-morrow*, knew them, and so did those rebuilding city housing estates after World War II. In cities internationally, these principles are reasserting themselves in medium-density infill housing and regeneration projects that prioritize community building, suggesting a more positive way forward and a revival of cities as family spaces. However, the problems of sprawl, isolation, the inheritance of poor high-rise buildings of the 1960s and 1970s, and the legacy of cities built for cars not people will not easily recede and will continue to blight children's lives unless planners, local government officers, housing and other environmental professionals prioritize resourcing the retrofitting of problem neighbourhoods. Consideration of children's needs and the inclusion of children in the remaking of city neighbourhoods are essential if the mistakes of the past are to be remedied.

When children in a Dunedin school were asked to identify what makes a good neighbourhood for children, the main themes were about connectedness (with nature and with neighbours) and safety from traffic, such as speed bumps

Table 5.4 Children's views on what makes a good neighbourhood

Speed bumps; colourful houses.	Some speed bumps so there is no racing.
A good neighbourhood is where you are very safe.	More parks with swings; internet, gyms, and slides.
Safety and speed bumps.	A great neighbourhood is safe.
The neighbour is caring and kind and says hello in the mornings and helps with chores like gardening.	Selling chocolates for fundraising*.
Having your own space and getting to know the people around you.	Getting chocolate!
Having neighbours that lend you stuff and are friendly and say hello.	All the park trees and space to think.
A quiet natural place to live in would make a great neighbourhood.	There's heaps of play area for kids.
A good neighbourhood is a place where you can feel safe and know a lot of people and a lot of space.	Because it's cool.
	A neighbourhood with big parks, swings, BMX areas and one long big slide!
	What makes a neighbourhood a good neighbourhood is size. But small neighbourhoods are good too.

Note: *In New Zealand selling chocolates in the neighbourhood is a common fundraising activity.

(see Table 5.4). Not only were they concerned about the physical space of the neighbourhood, but the social space was important to them as well. Children recognize the need for good services (playgrounds, libraries), quality environments (beaches) fun and excitement (sledding, pools), as is reflected in the neighbourhood proposed in the child's drawing and commentary in Figure 5.5. As for adults, children crave a good social and physical environment. Friendly people – or 'neighbourliness' – seemed to be as important as trees and space in which to think. They value space, but also 'belonging'.

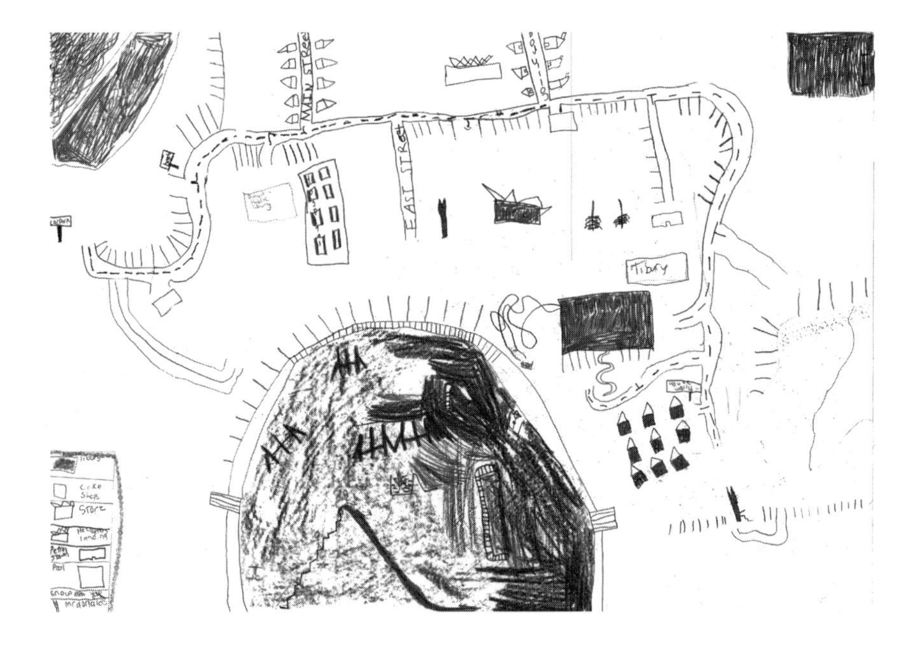

Figure 5.5 'If I was a town planner'

'If I was a town planner I would make sure some of the days were snowing so that we can go sledding. My town will have 12 libraries so people can read. There will be lots of shops where you get delicious cakes. There will be a huge playground for kids to play at. There will be 13 beautiful beaches. It will have nice clear, blue water so clear that you can see the fish. My parents are going to own a kids' clothes shop. There are going to be 200 helicopters and you can ride in them for free. My town has lots of pools for kids and hydro slides. You can go to outer space to see the planets. There is going to be lots of houses for people to live in. Next to the library is going to be a hospital. There is going to be a petrol station for cars to get petrol. There is going to be a pool with nice warm water.'

References

Alexander, C. (1966) 'A city is not a tree', *Design*, vol 206, pp46–55

Bartlett, S., Hart, R., Satterthwaite, D., de la Barra, X. and Missair, A. (1999) *Cities for Children: Children's Rights, Poverty and Urban Management*, Earthscan, London

Bassani, C. (2007) 'Five dimensions of social capital theory as they pertain to youth studies', *Journal of Youth Studies*, vol 10, no 1, pp17–34

Chawla, L. (ed) (2002) *Growing Up in an Urbanizing World*, UNESCO and Earthscan, London, pp111–134

Coleman, J. S. (1988) 'Social capital in the creation of human capital', *American Journal of Sociology*, vol 94, Supplement, ppS95–S120

Crow, G., Allan, G. and Summers, M. (2002) 'Neither busybodies nor nobodies: Managing proximity and distance in neighbourly relations', *Sociology*, vol 36, no 1, pp127–145

Driskell, D. (2002) *Creating Better Cities with Children and Youth: A Manual for Participation*, UNESCO, Earthscan, London

Freeman, C. (2010) 'Children's neighbourhoods, social centres to "*terra incognita*"', *Children's Geographies*, vol 8, no 2, pp157–176

Furedi, F. (2008) *Paranoid Parenting*, Continuum, London

Gill, T. (2007) *Can I play out...? Lessons from London Play's Home Zones Project Report*, London Play and London Councils, London, UK

Hall, T. and Vidén, S. (2005) 'The Million Homes Programme: A review of the great Swedish planning project', *Planning Perspectives*, vol 20, no 3, pp301–328

Howard, E. (1902) *Garden Cities of To-morrow*, S. Sonnenschein & Co., Ltd, London

Kearns, A. and Parkinson, M. (2001) 'The significance of neighbourhood.', *Urban Studies*, vol 38, no 12, pp2103–2110

Low, N., Gleeson, B., Green, R. and Radovic, D. (2005) *The Green City: Sustainable Homes Sustainable Suburbs*, UNSW Press, Sydney

Malone, K. and Hasluck, L. (2002) 'Australian youth', in L. Chawla (ed) *Growing Up in an Urbanizing World*, UNESCO and Earthscan, London, pp81–110

Offer, S. (2007) 'Children's role in generating social capital', *Social Forces*, vol 85, no 3, pp1125–1142

Pilch, T. (ed) (2006) *Neighbourliness*, The Smith Institute, London

Swart-Kruger, J. (2002) 'Children in a South African Squatter camp gain and lose a voice', in L. Chawla (ed) *Growing Up in an Urbanizing World*, UNESCO and Earthscan, London, pp111–134

Young, M. and Wilmot, P. (1957) *Family and Kinship in East London*, Routledge, London

UNESCO Growing up in Cities–Asia Pacific (2007) *How Friendly Is My Community: Children's Research Workshops Information Kit*, UNESCO Growing Up in Cities–Asia Pacific, Faculty of Education, University of Wollongong, Wollongong, Australia

6
City Centre

> In small towns I guess you can get really bored, but in London, there is always something to do. (Homeless girl, Focus Group)
>
> [There are] lots of things to do here. It's just a question of whether you want to do it or not, or if you have the money. (Girl at risk of offending, Focus Group)
>
> We only do things that are free, 'cause London is so expensive. (Homeless girl, Focus Group)
>
> My mum and dad always take us out and know lots of free places to go, but not many other parents seem to bother. (unattributed)
>
> – *the quotations above are from* Making London Better for All Children and Young People, *Mayor of London (2004)*

City Centres Matter

The city centre is the heart of the city: it defines the city and sets the tone for the city. If the city centre is vibrant and successful, it invigorates the rest of the city, and if the centre is deserted and broken down, the sense of decay spreads and permeates the broader city psyche. So, too, city centres that are open and welcoming of children, where children are clearly visible, and that have children's welfare at the heart, spread this sense of children's centrality out beyond the city core into the neighbourhoods and into the wider public life of the city. Children's visibility matters because the city centre is also the political nucleus. It is where the laws, policies and decisions are made that infuse the life of its citizens for better or worse. If children are to be fully valued in the city, it is vital that they are present at its core, and that the city centre in its approach to children positively reflects and supports children's rights to be there and their right to experience the social, economic and historic wealth that is concentrated in the city centre.

City centres that welcome children spread children's centrality into the life of the city

While the physical design of the city is important, a good physical environment alone is not enough. Social and cultural factors determine if a city works for children. The social and cultural values present in the city, in its people, designers, managers and decision-makers, influence whether children are able to relax and play in the city plaza, whether young people are welcomed into or discouraged from its shopping centres, whether the city centre facilitates access for people or just for cars, and whether the child feels safe or threatened in the city. City centres differ markedly in the extent to which they welcome and cater for children. This chapter examines the city centre as a place for children: its changing nature, its challenges, its public spaces, its role as a place where children live and as a dynamic focus of urban redevelopment.

City-Centre Functions

The city centre represents a coming together of the social, cultural, recreational, educational, political and economic functions that provide the lifeblood of the city. Children use the city centre in diverse ways; they can physically live in the city, but more often frequent the city to use its facilities. Table 6.1 summarizes some of the key functions and elements found in city centres. Usually, the larger the city, the greater is the range of functions. For example, most cities will have a hospital; but only the largest cities will have specialist medical hospitals such

Table 6.1 City-centre facilities (those most likely to be used by children are italicized)

Economy	*Shops*
	Markets
	Eateries, restaurants and cafés
	Pubs
	Offices
	Industries
Public services	*Public service outlets (e.g. welfare offices)*
	Hospitals and specialist medical services
	Rest rooms and toilets
	Central and local government offices
	Centres of justice (law courts)
	Schools
	Universities and tertiary institutions
Cultural	*Museums*
	Festivals, celebrations, parades
	Sports centres
	Religious and spiritual places
	Libraries, theatres and dance centres
	Art galleries and public art
	City quarters (e.g. Chinatown)
Transport	*Railway stations*
	Trams and bus interchange
	Pedestrian walkways and footpaths
	Cycle ways
	Ports and river/sea ferry terminals
	Major road interchanges and networks
	Bridges
Built environment	*Houses and apartments*
	Dedicated child spaces, childcare centres and playgrounds
	City squares/plazas
	Monuments and statues
	Fountains
	Heritage buildings
Natural environment	*Public parks*
	Gardens
	Street trees

as children's/paediatric hospitals, burns hospitals, spinal injury units, and eye hospitals. Hospitals are especially worthy of consideration as they are visited by children who are vulnerable and sick; thus, good accessibility, parking and public transport, places to play and relax around visits, and places to eat become especially pertinent. For children visiting the hospital, services and access provision are especially significant, as in: do train or bus stations serving the hospital have lifts; ramps; clean, appealing and warm waiting areas; and wheelchair-accessible toilets? The facilities that children are most likely to use in the city, and which need special consideration from planners and urban managers, are highlighted in Table 6.1. However, as the hospital example indicates, the presence of support facilities and services can be important too. The lack of emphasis for certain variables in Table 6.1 does not indicate a lack of importance; for instance, while centres of justice have not been italicized in the table as 'most likely to be used', there are times when children may need to access law courts/family courts. These are often critical times in a child's life when they may be especially vulnerable. It is crucial that children who come into the court setting are supported during the time spent there. The presence of myriad city-centre facilities – museums, theatres, railway stations, shopping arcades, public gardens – provide children with immense opportunities to participate and observe the breadth of public life. As cities spread out, children confined to the suburbs can become isolated from this public life, a process Lennard and Crowhurst Lennard (1992) describe as the '*de facto* amputation' of children from the world of the city. If the city is to work for children, it needs to be available and welcoming to all children regardless of where they live.

City Centres: Legacies and Challenges

Tourism can be a great indicator of which cities work. Tourist magnet city centres, such as Venice, Prague, Vienna, Edinburgh and Copenhagen, are characterized by easy public access; interesting high-quality architecture; a sense of heritage; parks and squares that abound with public art and which encourage relaxation, eating, sitting, chatting and 'people watching'. It is the older cities, with their human scale, their mixed and diverse uses, the pedestrian streets, eateries and on-street human life, which attract people. It isn't hard to know whether a city centre works: it feels 'full', it has a sense of happening, it abounds with life and activity. The city centre is the powerhouse of urban life as Lewis Mumford, a 'brilliant critic of architecture and society', wrote in 1938:

It isn't hard to know whether a city centre works

> The city as one finds it in history is the point of maximum concentration for the power and culture of a community. It is the place where the diffused rays of many separate beams of life fall into focus ... The city is the form and symbol of an integrated relationship: it is the seat of the temple, the market, the ball of justice, the academy of learning. Here in the city the goods of civilization are multiplied and manifolded; here is where human experience is transformed into viable signs, symbols, patterns of conduct, and systems of order. (Mumford, 2004, p16)

In 1938, Mumford believed that cities were in crisis and argued the need to act against the destructive forces that would leave cities 'blasted and deserted', and as little more than 'cemeteries for the dead'. His negative but also prophetic words were echoed by many later planners and architects who became concerned with the dehumanizing scale of city development. Cities, in the latter half of the 20th century, especially those in the US, Australasia, Japan and the UK, became dominated by high-rise office blocks and motorway routes and intersections. People moved out, taking with them the doctors, corner shops, street life, schools, community halls, pubs, cafés, cinemas and the minutiae of daily life. Not only were people leaving, but the economic heart of the city experienced challenges as ports and industries closed and commerce moved out to the edge of the city or even to different countries. Left behind were deserted city centres. In her book *The Death and Life of Great American Cities* (Jacobs, 1961), written at the height of the urban crisis, Jane Jacobs examined what made cities work and emphasized the need to recognize what she saw as the true essence of city life – people:

Jane Jacobs recognized the true essence of city life – people

> This ubiquitous principle is the need of cities for a most intricate and close-grained diversity of uses that give each other constant mutual support, both economically and socially. (Jacobs, 2004, p34)

Fortunately, not all cities have experienced decline, with Jacobs's 'ubiquitous principle' of peopled diversity remaining intact for typically older European and Asian cities.

For some cities, though, the process of decay continues with the decentralization of city-centre shops (into out-of-town shopping malls), cinemas (into suburban multiplexes), jobs (to business parks) and services all migrating outwards, taking with them the city core's reasons for being. Children are particularly affected by this decentralization process as it spreads activities over a wider area, and diminishes ease of access for children along with the elderly, disabled and others who are reliant on walking and public transport. Decentralization of shopping from public streets into malls has particular impacts upon children and young people. Shopping malls are usually private spaces in which access can be strictly controlled. Children and young people may find themselves targeted by private security firms who perceive them to be a threat if they do not appear to be participating in primary consumption activities, but just 'hanging about'. In such cases, the children and young people are 'encouraged' to 'move on' and out of these privatized spaces (White, 1993, 1996; Hil and Bessant, 1999). As shops, offices, leisure, eating and other activities migrate out, so do people and so, too, does the soul of the city, leaving behind empty buildings and streets, and disused parks, which, in turn, leaves the city centre feeling less safe and usable, continuing the spiral of decline.

Across the globe in cities as diverse as Manchester, Sydney, Boston, Rotterdam and Gothenburg, the decline of the city centre has galvanized governments into a process of regeneration and renewal, and a drive to repopulate the city. City managers recognize that the old model of a city centre dependent upon industry and supported by a dense network of working-class homes is gone and a new city form still with people at its centre is necessary. Recreating the city for people becomes imperative, ushering in an era of redevelopment programmes designed to bring residents, including families, back to the centre. In 2006, research commissioned by the Belfast City Council looked at how the city could revive its city centre and join cities such as New York or Vancouver in embracing city living with a vibrant, safe and successful 24-hour street life (Belfast City Council, 2006, p5). One of the key components identified as vital in achieving success for Belfast was:

> Family-friendly residential development: Strict planning policies must be jointly balanced by family-friendly development within the city-urban housing. Belfast must match the needs of families with children in ways that can match the attraction of suburban housing estates. (Belfast City Council, 2006, p4)

Globally, cities are seeking to attract people, including children and families, back into the city. Certainly, repopulation of the city centre is an excellent goal; whether city centres are good places for children is less certain. It is in the public spaces that the city as the focal centre of public life is most likely to be experienced by children.

City Centres as Public Space

Public space is where children experience the diverse social and cultural life of the city. These spaces present a valuable counterpoint to the privatized, controlled and regulated places that constitute much of the child's daily world. Public spaces are observation points where children see how the world works: they teach children about the society in which they live (Lennard and Crowhurst Lennard, 1992), and provide contact points and places of encounter where children meet others who are similar and different from themselves:

Public spaces are where children see how the world works

> Public space is where the drama of life unfolds. The streets, squares and parks of the city give form to the ebb and flow of human exchange. These dynamic spaces are an essential counterpart to the more settled places and routines of work and home, providing channels for movement, the nodes of communication, and the common grounds for play and relaxation. (Carr et al, cited in Peattie, 1998, p249)

Traditional cities recognized and valued these needs, endowing cities with piazzas, market places, city squares, gardens and parks where social gatherings could happen spontaneously or by arrangement. In more recently designed towns and cities, such spaces are less apparent. In the enclosed shopping centres and shopping malls that characterize many modern town centres, gathering spaces are dominated by eateries who welcome only paying food consumers to use their seats and tables; the outside space is dominated by car parks, gardens give way to sterile grass-dominated landscapes, and there are few sitting and 'hanging' out places. Some malls deliberately design out sitting places to 'encourage' sitting in commercial food places and to discourage young people from 'hanging about'. More positively, some cities have reclaimed and enhanced public space, pedestrianizing streets, providing sitting places, and encouraging street entertainment, buskers and street stalls (see Figure 6.1). Where streets have been pedestrianized, the initial objections of the commercial sector to removing cars from the streets invariably subside as the economic benefits of having people staying longer in the city centre are realized.

Figure 6.1 Buskers in public spaces make an important contribution to the cultural life of the city and are enjoyed by young and old alike (Bethesda Terrace, Central Park, New York City)

Source: Paul Tranter

For children and their parents, city centres can be a fraught place; both get tired and need time out from the hustle and bustle of the commercial sector. Public sitting places meet this need. Effective places for 'time out' can be small places. In Wellington, New Zealand's capital city, a popular hanging out place for families is at the Buckets water sculpture, where water cascades from the top bucket, eventually tipping the lower buckets with noisy splashes (see Figure 6.2). The sculpture is in a pedestrianized street surrounded by cafés and shops,

Figure 6.2 The Buckets are a well-loved feature in the cityscape of Wellington, New Zealand's capital city

Source: Claire Freeman

an area that also acts as a meeting place for young people. Public spaces can be widely divergent, ranging from the vast squares such as Red Square in Moscow, to the heritage squares such as St Peter's Square in Rome, to the small intricate places such as Wellington's Buckets, known only to locals. Although public spaces may be diverse, the characteristics that can make a space successful are common. Successful places:

- are easily accessible and can be seen by potential users;
- clearly convey the message that they are available and meant to be used;

- are beautiful and engaging;
- are furnished to support the most likely and desirable activities;
- provide a feeling of safety and security;
- offer relief from urban stress and enhance the health and emotional well-being of its users;
- are geared to the needs of the user group most likely to use the space;
- encourage use by different groups;
- are accessible to children and people with disabilities;
- provide an environment that is physiologically comfortable;
- incorporate components that can be manipulated (adapted from Cooper Marcus and Francis, 1998).

Good public space for children does not need child-oriented play equipment

Although this list was not compiled specifically for child users, the criteria, if applied, would enhance spaces for children. There are, nonetheless, some features that make spaces particularly appropriate for children, such as changes in levels; informal as well as formal seating; space for running and active play; intimate spaces; places to watch entertainment; variety; and interesting features, such as statues, fountains, steps, live action and the presence of other children. Contrary to the pervasive thinking on the part of urban designers and managers, good public space for children does not need to incorporate child-oriented play equipment. The Wellington Buckets are not a child's play structure, but nevertheless provide fun, enjoyment and entertainment for all ages.

Making Public Spaces Work

Public space presents a particular challenge to city managers as it brings together people and activities that may conflict. The classic in this regard is public space that attracts both the elderly and skateboarders. While the incompatibility between more sedentary elderly people and more active skateboarders is somewhat understandable, less so is the fact that some adults are intolerant even of the presence of children, and especially the presence of young people. In a *Play England* study, researchers reported how 'the police described receiving complaints from adults about children who, although not doing anything illegal, were irritating or alarming them by gathering in groups'. The report concludes from this that there was 'an underlying tacit – or overt – disapproval from adults of the everyday presence of young people in public' (Beunderman et al, 2007, p76). These intolerant attitudes can lead to the exclusion and alienation of children from public space. One response from children and young people is to avoid public places; another is to colonize the 'leftover spaces': places on the margins, the less popular places away from adults (see Figure 6.3).

Children as well as adults can feel threatened and afraid in public spaces. A very large study of the experience of British 10- to 12-year-olds which asked whether they liked their city centres found that most liked their city: 42 per cent said it was OK, 37 per cent like it, 12 per cent love it, and only 8 per cent dislike or hate it (Woolley et al, 1999a). Despite this overall positive response, fear was still a real issue, with one third of the children describing the city centre as dangerous and one fifth describing it as violent. These fears were generated in a number of ways, including from subliminal messages associated with run-down

Figure 6.3 Colonizing leftover space, under the Arches, South Bank, London

Source: Claire Freeman

and degraded areas, subways, backstreets, dark or poorly lit areas, and the presence of 'worrisome' people, including uniformed security guards, beggars, street sellers (as well as homeless people selling the *Big Issue* magazine), groups of teenagers, gangs of children, 'druggies' and 'Goths'. The sense of unease seemed to be higher for children who lived out of town, suggesting that children more familiar with the city-centre environment learned how to relate to it, exhibiting higher levels of what Cahill (2000) terms 'street literacy'. On the positive side, children in this study were frequent users of the city centre, with over 70 per cent visiting it weekly (Woolley et al, 1999b). The children also exhibited strong 'civic values' in that they disliked litter, dirt, smells, pollution, disorder and incivilities, and liked the presence of closed circuit television (CCTV), the presence of police, and appreciated city-centre enhancement programmes that provided public art, seating, fountains and general improvements (Woolley et al, 1999b). It is important to recognize the need to separate incompatible uses, such as spaces for the frail and vulnerable, and more active spaces for older children and cyclists when designing public space. In most cases, though, children's needs (good services and accessibility), their dislikes (dirt, neglect, gangs, sterility and fear) and their likes (aesthetics, street entertainment, safety and conviviality) are factors that meet with common public accord. One feature that has been prominent in recreating good public spaces and a central focus of regeneration initiatives has been to furnish the spaces with public art, primarily sculptures, statues and fountains.

Larger spaces such as Federation Square in Melbourne, Australia, show that by using varied architecture, landscaping and spatial arrangements, diverse needs can be provided within a single city square. As part of its plan to attract

Figure 6.4 Children gravitate towards public art that encourages interaction, as with these 'sticks' which make a noise when shaken (Federation Square, Melbourne, Australia)

Note: The guard around them, however, suggests that interaction with this artwork is not intended.

Source: Otago University Planning Programme

people back into the city, Melbourne has created excellent public spaces. Pavements have been deliberately widened to facilitate street life. Public art has been strategically placed around the city. The centrepiece is Federation Square, which, when opened in 2002, was highly controversial due to its cost (AU$420 million) and modern architecture (Fed Square Pty Ltd, 2009). It comprises a range of interestingly, some say outrageously, designed buildings, and includes covered and open spaces, links to the Yarra River and a new park. Although not designed for children, it nonetheless works for children. The big screen is used to broadcast major events; the floor design, with its changing surface materials, low walls and changes in level, provides opportunities to hang out, play, climb, run and play ball. In 2008 it hosted 2323 events in its public spaces, of which 87 per cent were free and open to the public. The square is now deemed highly successful, attracting 8.41 million visitors during 2008 to 2009. Many of these are children, who will use the square and its public art in ways not necessarily intended by the designers (see Figure 6.4).

Public Art

In public art lies an opportunity to engage children directly with the cultural life world of the city. The city centre represents the main gathering place for public art, which has long been a famous focal feature for public life in Italian cities such as Rome, Florence, Milan and Venice. More recently, public art has been a major component in the redevelopment strategies of cities such as Belfast (Northern Ireland), Birmingham (UK), Seattle (US), Melbourne (Australia) and Vancouver (Canada). Public art presents an opportunity for children to engage physically and actively with art and to be engaged in the creative process. Public art meets one of the fundamental requirements of a child-friendly city: it is 'free' and everyone can engage with it. Public art relates to children in a number of ways.

Appreciative art: Art to be admired and art as history

Public art provides a way for children to engage with the history and identity of a place. The Big Fish in Belfast is made up of blue tiles that tell the history of the city. The Bull sculpture in Birmingham's Bull Ring Shopping Centre harks back to its history as a farmers' market (see Figure 6.5), and Copenhagen's Little Mermaid links to the Danish folk tale. The great historical sculptures of Rome and Paris also create historic connections. Thanks to modern films, children now more easily connect to the gargoyles of Notre Dame that featured in Disney's film *The Hunchback of Notre Dame*.

Fun art: Art that creates a sense of playfulness and joy

Public art can appeal to children's sense of fun and provides important markers for children visiting the city centre: Adelaide's (Australia) life-size bronze pigs (Horatio, Truffles, Augusta and Oliver) rooting around in the litter bin; the pig on a diving board in Prague; and the fun sculptures in the fountain by the Pompidou Centre in Paris. Impermanent art such as 'living statues' also appeals to children.

Figure 6.5 The Bull in the Bull Ring Shopping Centre is an imposing and historically relevant piece of public art (Birmingham, UK)

Source: Claire Freeman

Art as interaction and adventure

Public art appeals to children's sense of adventure: to touch it and to climb it. Some of the public art installations associated with the redevelopment of city centres are wonderful play opportunities for children. The water features from Sydney's harbour-side development (Figure 1.2 in Chapter 1) are excellent examples, as is the water feature in the Custom House Square in central Belfast, with its 84 jumping water jets. The relationship between adventure, interaction and art can become problematic. Since their erection in 1868, few statues have been more climbed than the Lions in Trafalgar Square (see Figure 6.6). Most children from the age of about seven will attempt (usually with some help) to climb the Lions (try googling 'Lions Trafalgar Square images' and see how many show people sitting on the lions, including many family photos). Climbing 'artworks' can bring children into conflict with the authorities and other members of the public. A recent example of this are the 15 wonderful Steam

Figure 6.6 The Lions in Trafalgar Square, London, have acted as a climbing magnet for generations of children

Source: Robin Quigg

Figure 6.7 The Steam Sculptures beg to be climbed, with their enticingly positioned footholds that evoke those found on climbing walls (Brisbane City Square, Australia)

Source: Claire Freeman

Sculptures that are dotted about Brisbane City Square in Australia. These are large round spheres, made from 7000 metal steamers and 780 plates bolted together. The spheres are different sizes. The Steamers provide excellent foot and hand holds for climbing and they prove irresistible to children and some adults (see Figure 6.7). Yet, climbing them is frowned upon and creates tension for parents and their children who recognize their climbability but also that to do so may be disapproved of. Placing appealing, climbable public artworks that beg to be explored in public places that children frequent, and then surrounding these with a no touch policy, can be a source of immense frustration for children and their parents.

Art as process: Children take a role in creating the art

Children have become involved in the production of many artworks created as part of city-centre redevelopment strategies, either through work with artists in schools or with community groups. Placing artwork that children have contributed to in important public spaces can be an immensely validating process (see Figure 6.8). Examples of public installations, some of which involved working with children and young people, can be seen at www.arteofchange.com/content/weaving-identities#london_art.

Public art provides opportunities to be dual purpose; it can have artistic appeal and also act as a play feature. Rather than provide the somewhat unappealing multicoloured tubular steel generic play equipment for children in the city centre, interactive public art installations can be used to provide aesthetically pleasing child-friendly living spaces that have multi-age appeal.

Figure 6.8 Children's designs incorporated within a high-quality artistic, yet functional, tree protector (Wollongong, Australia)

Source: Claire Freeman

City Centres as Home: Living 'Downtown' in the City

While city centres are places for children to visit, for other children the city centre is also where they live. To establish what is happening to children in the inner city is difficult as a number of simultaneous and conflicting trends are occurring. The overriding trend is a decline in the number of children and families living in or close to the city centre, and an associated decline in affordable family housing that pushes children still living there into often inappropriate high-density vertical apartments. In contrast, there is a trend towards encouraging repopulation of the inner city through specific development initiatives designed to encourage family living. In looking at children in the city centre, we argue that cities can be excellent places for children to live in, but that their service, play and housing needs must be provided for.

Overall, the proportion of people living in the inner city with children is probably lower than at any other period historically. Low numbers means that children may become a minority population whose needs can easily be – and are being – overlooked in favour of majority population groups. The decline in inner-city living has been especially prevalent in the larger industrialized cities. In the UK, for example, between 1961 and 1991 the central area population in Manchester fell by 73 per cent from 4000 to 1093. In nearby Liverpool city centre, Abercrombie and Everton wards (districts) experienced population losses of 44 and 64 per cent, respectively, between 1971 and 1991. In a study of 45 US cities, however, there was an 8 per cent increase in housing units between 1970 and 2000 in the city; this compared to a 99.7 per cent increase for the suburbs (Birch, 2006). The population decline in city centres is caused by rising unaffordability, lack of investment in city-centre facilities for families (e.g. for health clinics and schools), perceptions of city centres as areas of decline and deprivation, safety fears, and an absence of attractive features seen to be offered in suburbs and smaller cities.

Contrary to popular perception, however, children can be found living in high-density inner-city housing. Although there has been no statutory regulation prohibiting children living in inner-city high-density developments, the view prevalent among developers is that children do not belong in such developments. Their intended market comprises singles and child-free young or 'empty nester' couples (Costello, 2005). In a study of developer attitudes in Melbourne, Australia, Fincher (2004, p337) became concerned by developers':

> ... presentation of children ... as burdensome – people from whom to 'escape' and 'be freed' ... the construction of a myth that the perfect lifestyle is to be lived without children or other dependants – one's own or other people's.

This argument has also been used to promote child-free gated communities, some condominiums in the US and adult tourism developments, and is implicit in elderly people's residential complexes even though children frequently visit relatives there. Waterfronts and docklands have been a particular focus for housing development initiatives that reflect this 'child-free' stance. However, the fact that buildings are built for 'non-families' does not mean that families don't live in them. Children are, indeed, living in dense inner-city apartments, many of which do not meet their space and play needs (see Figure 3.1 in Chapter 3). Across Australia, developers are building apartments that fail in this way; yet census data showed that by 2001, 20 per cent of people in Australia living in high-rise units were families, and most such units would be in or close to the central business district (CBD) (Mizrachi and Whitzman, 2009). In Melbourne, by 2021, approximately 4000 children under the age of 12 will live in the inner Melbourne areas of the CBD, Docklands and Southbank.

Waterfront and dockland developments have been a particular focus for a 'child-free' stance

What, then, is it like for children living in the inner city – or 'downtown', to use the American term? There are advantages for children related to living in the city centre. Parents are near their places of work, so children benefit from

Children gain from the diversity associated with city centres

the extra time gained from their parents not needing a long commute. Inner cities can offer excellent parks, often superior to park provision in the suburbs. Children gain from the diversity: the cultural and social mix associated with city centres, its ethnic quarters with their festivals, eateries and different ways of interacting. But there are also limitations: family housing is in short supply and prohibitively expensive; doctors, schools, supermarkets and other primary services that families need can be scarce; and children can become isolated in high-density apartment blocks. Where inner-city living has been a positive choice for families, it can be beneficial; but for families forced by poverty or other lack of choices to live with their children in high-density or cramped housing, the experience can be isolating, stressful and prejudicial to children's well-being (Mizrachi and Whitzman, 2009). In Tokyo, Yabe (2003) echoes these concerns, finding that families would move to the inner city but for a lack of public or private affordable housing and a lack of services such as early year childcare. Ongoing gentrification of family housing stock in the inner city and house prices unaffordable for other than very high-income families mean that the decline is likely to continue. It is not inner-city and city-centre living in itself that is counter to children's needs, but the lack of affordability and suitable housing that is the main problem.

Many cities have become concerned about population drift and associated city-centre decline. These cities seek to reverse counter-urbanization trends, to create vibrant city centres as places for people to live and work in. As part of their regeneration strategies, family housing is being erected on land left vacant by departing industries or the removal of derelict buildings. Since the late 1980s, there have been concerted efforts on the part of cities to redress this population decline and to attract residential populations back into the city. Melbourne's city-centre population grew by 18 per cent per annum between 2001 and 2007, and now comprises 20 per cent of the city's population (City of Melbourne, 2008). Similarly, some 40 per cent of the population growth in Vancouver between 1971 and 2001 took place in the metro core (City of Vancouver, 2005). However, care has to be taken in equating general population growth in the city centre with an increase in families. Very few families with children were found in a recent study to be living in the new sustainable city-centre housing projects initiated by local governments in the large UK cities of Bristol and Swansea (Bromley et al, 2005). One example of a successful initiative to provide for inner-city living for families is the Coin Street housing project in London. The initiative shows the role and value of affordable family social housing in redressing population decline.

Building Success in the City Centre: Coin Street

The Coin Street development covers an area of some 35ha and is located in London, close to the nation's cultural and political heart. Situated on the South Bank of the River Thames in London, Coin Street is a mere stone's throw from the Houses of Parliament, Westminster Abbey and international tourist attractions such as St Paul's Cathedral and the Tate Modern Art Gallery, and is adjacent to the South Bank theatre complex and riverside attractions such as the

London Eye (a huge observation wheel). The site has been developed by the non-profit group Coin Street Community Builders. In 2006 it reported that '220 high quality, well designed homes at affordable rents for people in housing need have been developed to-date ... a mix of 1, 2 and 3 bedroom flats, 3, 4 and 5 bedroom houses ... run by four fully mutual housing co-operatives' (CSCB, 2010). The intention is to eventually provide 600 homes. In addition to housing, the development includes the very successful Oxo Building, comprising a restaurant, commercial premises, 78 cooperative flats, and Gabriel's Wharf and Market (CSCB, 2010). The development, which began in 1984, was a groundbreaking departure for a site with such high real-estate value and quite a different proposal for the site from the high-tech, high-rise commercial development proposal put forward by internationally renowned architect Richard Rogers.[1]

The housing has been designed to suit families, with most housing having small private gardens, yards and balconies and opening onto inner courtyards with communal green space. With the development of family homes has come the need to provide services that have moved away: childcare, family support centres and affordable meeting places. To this end, a neighbourhood centre was completed in 2007 (Haworth, 2009). To reiterate the earlier question, 'What is it like for a child growing up in the inner city?', Table 6.2 lists just some city-centre features and facilities that a child from Coin Street area could access in their inner London local area. Overall, the area is diverse, vibrant, multicultural, has good to exceptional public transport, as well as excellent cultural facilities and public spaces. How much of this cultural and other wealth, though, would realistically be accessible to the child, especially given the predominance of low incomes among Coin Street families? While superb leisure facilities such as the London Eye, national theatres, world-class restaurants and markets such as Covent Gardens are certainly present, it is unlikely that children would, in practice, use these other than on a very occasional basis. Free facilities such as the Tate Modern Art Gallery, various parks and gardens, and free events such as the 'changing of the guards' outside Buckingham Palace are more likely to be accessed and used. The most used places are likely to be the free public spaces, Gabriel's Wharf and the riverfront closest to where children live. Further afield are other museums, markets, shops, sports centres and educational institutions that children can benefit from, and most museums have free access. A major contributor to accessing these and other further places is the free bus and tram travel for children up to the age of 15 using the Oyster card, with some free travel on trains for adult-accompanied children, with older children travelling at the child rate. Free and reduced-cost travel on public transport is also essential for children living outside the city centre to access the kind of free cultural and other events and activities available to children living in the city centre.

Life in the city centre can be a challenge for children and their families

Life in the city centre can be a challenge for children and their families, especially where cost precludes accessing adequate housing, access to theatres, art galleries and sports centres; but these challenges can be addressed, if not always mitigated, where a commitment is made – as at Coin Street – to developing viable inner-city communities.

Table 6.2 Features and facilities close to Coin Street

Cultural places	Transport and bridges	Parks and gardens	Embankment/Riverside	Government buildings and other services	Squares/meeting places	Tourist attractions
• **Tate Modern** • London Museum • *London IMAX* • *Young Vic Theatre* • *National Theatre* • *Royal Festival Hall* • *Globe Theatre* • *London Aquarium* • *National Film Theatre* • *London Warehouse Recording Studio*	• **Waterloo Station** • **Tube stations** • Hungerford Bridge • Millennium Bridge • Waterloo Bridge • Southwark Bridge • Blackfriars Bridge	• **Many gardens around Coin Street** • **Bernie Spain Gardens** • Jubilee Gardens • Victoria Embankment Gardens • Whitehall Gardens	• **Gabriel's Wharf and Market** • **Riverside walkway** • Oxo Tower Complex • *London Eye*	• **Waterloo Library** • **Johanna Primary School** • **Blackfriars Medical Practice** • **Sainsbury's supermarket** • **Eateries** • Pubs	• Somerset House fountains • Trafalgar Square • Covent Gardens	• Various monuments • *Buckingham Palace* • *Houses of Parliament* • *Westminster Abbey* • *St Paul's Cathedral*

Note: boldface = most likely to use; usually free entry; italics = restricted; usually paid entry.

Children's Views on What Makes a City Centre a Good Place

Primary school children in Canberra were asked to identify the things that make a city centre a good place for children. A number of themes emerged that reinforce the arguments presented already in this chapter, with repeated suggestions for more seats, toilets, jobs for children and cultural food (from different countries) (see Table 6.3). As for the school grounds (see Chapter 4), there was also an identification of the need for more bubblers (drinking fountains), as well as mention of more plant life and gardens – indicating a desire for more nature in the city centre. Figure 6.9 shows a drawing by a fifth-year child of a city centre as a good place for children. The drawing includes trees and grass, as well as a youth centre, medical centre, food shops and children's playground.

Affirming the Role of City Centres

Cities can and should be exciting and interesting places for children. With many cities showing increasing numbers of children living in the city centre, focusing

Table 6.3 Children's views on what makes a good city centre

Big parks with playgrounds, mazes, sculptures and lots of amazing old buildings and water fountains, like in Europe; lots of concerts.	An easy way to navigate around the mall and find your favourite shops; should be a map or something to find the right place. Also it should be bright and happy, not smelling bad and dark; no really loud music you can hear a mile away.
Kids' play centre; more exit signs.	More toilets and change rooms for baby and toddlers.
More cultural food shops like Mexican, Brazilian and Indian; more gaming stores or hobby stores.	More kids places; less adult shops; you can drop off your kids if you want to do proper shopping; more book stores.
Sports gear; less fast food because it's too busy and loud.	More fresh air, not just air conditioning and plant life.
People that are happy; video game shops.	A youth centre, youth help and support.
More comfortable seats, more cultural food, more gardens.	More fun places for kids, ban smokers, more seats, places to relax, wheel chairs for old people that need them.
I would like to have a no smoking city centres policy because I hate smokers.	They need to add more seats so you don't have to walk all the way down when you're tired.
More seats and escalators that go across the floor so you don't have to walk.	I wish more shops could have little jobs for kids our age and more parking and more places for people to sit down.
There should be more toilets so we don't have to run really far just to go to the toilet.	A little section of a shop for kids (e.g. craft shop, a little corner with craft stuff that kids could use).
A safe place where we could stay; a place where kids could work.	Clean and more toilets around; bubblers if you get thirsty, cultural food stores, more electronic and hobby stores like EB games.
More food stalls around the mall, more comfortable chairs.	More toilets, not too many of one shop.
Some sick bays, more fun places, jobs for kids.	
More good shops, more shops for kids; shops where 11-year-olds can get jobs.	

Figure 6.9 Drawing by Canberra school child of the city centre as a good place for children

Source: Year 4/5 student at Blue Gum School, Canberra

on children's needs is imperative. Living conditions for children in the city are highly variable, with accommodation ranging from excellent purpose-built family housing, as at Coin Street, to children living in unsuitable accommodation due to poverty and lack of affordable accommodation choices. The prevailing view that city centres are not appropriate child spaces needs to be challenged, especially as it relates to family living and public space. For cities to work for children and their families there needs to be provision of services, from schools to toilets, and affordable entry to places of art, culture, history, learning and eateries. Accessibility is imperative and must include good affordable public transport for children travelling independently or with their families. In much of the developed world, children's independent access to the city centre has decreased as part of a general decrease in independent mobility, but also as a consequence of perceptions of lack of safety associated with city centres, especially at night. Recent years have seen some excellent city-centre redevelopment strategies, including waterfront developments, pedestrianization, revival of city squares, public transport enhancement, and initiatives to repopulate the city and its streets. The trend towards out-of-town shopping centres and shopping malls reduces children's ability to engage with civic life, culture and history traditionally associated with city centres and should be challenged. Children need to be seen in city centres to remind decision-makers that they have to consider children in this central part of their living environment. The revitalization initiatives that cities are undertaking reaffirm and enhance the city's role as the core of public life for all of its citizens, including children.

Note

1 Famous buildings on which Rogers has worked include the Pompidou Centre in Paris, the Lloyds Building and the Millennium Dome in London, and Marseilles, Madrid and Heathrow airport terminals.

References

Belfast City Council (2006) *Belfast State of the City: The Development Brief, October 2006 – Population Change,* Belfast City Council, Belfast

Beunderman, J., Hannon, C. and Bradwell, P. (2007) *Seen and Heard Reclaiming the Public Realm with Children and Young People*, Play England, London

Birch, E. L. (2006) 'Who lives downtown?', in A. Bérubé, B. Katz and R. E. Lang (eds) *Redefining Urban and Suburban America Evidence from the Census 2000*, Brookings Institute, Washington, DC

Bromley, R. D. F., Tallon, A. R. and Thomas, C. J. (2005) 'City centre regeneration through residential development: Contributing to sustainability', *Urban Studies*, vol 42, no 13, pp2407–2429

Cahill, C. (2000) 'Street literacy: Urban teenagers' strategies for negotiating their neighbourhood', *Journal of Youth Studies*, vol 3, no 3, pp251–277

City of Melbourne (2008) *Small Area Economic and Demographic Profile*, www.melbourne.vic.gov.au, July 2008, accessed March 2010

City of Vancouver (2005) 'Population and job growth in the metro core, city and region', *Economy – Structure, Info Sheet 1: Economy – Population and Job Growth, Metro Core, City, Region*, 9 November 2005, www.vancouver.ca/commsvcs/planning/corejobs/pdf/research/13popjob growth.pdf

Cooper Marcus, C. and Francis, C (1998) *People Places: Design Guidelines for Open Spaces*, John Wiley, New York, NY

Costello, L. (2005) 'From prisons to penthouses: The changing images of high-rise living in Melbourne', *Housing Studies*, vol 20, no 1, pp49–62

CSCB (Coin Street Community Builders) (2010) *A Very Social Enterprise*, www.coinstreet.org/housing/housing.html, accessed March 2010

Fed Square Pty Ltd (2009) *Annual Report for the Year Ended 30 June 2009*, Fed Square Pty Ltd, www.federationsquare.com.au

Fincher, R. (2004) 'Gender and life course in the narratives of Melbourne's high-rise housing developers', *Australian Geographical Studies*, vol 42, no 3, pp325–338

Haworth, G. (2009) 'Coin Street housing: The architecture of engagement', in A. Ritchie and R. Thomas (eds) *Sustainable Urban Design an Environmental Approach*, 2nd edition, Taylor and Francis, Oxon

Hil, R. and Bessant, J. (1999) 'Spaced-out? Young people's agency, resistance and public space', *Urban Policy and Research*, vol 17, no 1, pp41–49

Jacobs, J. (2004) 'Orthodox planning and the north end', from the Introduction to *The Death and Life of Great American Cities* (1961) in S. Wheeler and T. Beatley (eds) *The Sustainable Development Reader*, Routledge London, pp30–34

Lennard, H. L. and Crowhurst Lennard, S. H. (1992) 'Children in public places: Some lessons from European cities', *Children's Environments*, vol 9, no 2, pp37–47

Mayor of London (2004) *Making London Better for All Children and Young People: The Mayor's Children and Young People's Strategy*, Greater London Authority, London, January

Mizrachi, D. and Whitzman, C. (2009) 'Vertical living kids: Creating supportive environments for children in Melbourne central city high rises', Paper presented to the State of Australian Cities Conference, University of Western Australia, Australia, November 2009

Mumford, L. (2004) 'Cities and the crisis of civilization', in S. Wheeler and T. Beatley (eds) *The Sustainable Development Reader*, Routledge London, pp15–19

Peattie, L. (1998) 'Convivial cities', in M. Douglas and J. Friedman (eds) *Cities for Citizens: Planning and the Rise of Civil Society in a Global Age*, John Wiley and Sons, Chichester, pp247–253

White, R. (1993) 'Young people and the policing of community space', *Australian Journal of Criminology*, vol 26, pp207–218

White, R. (1996) 'No-go in the fortress city: Young people, inequality and space', *Urban Policy and Research*, vol 14, no 1, pp37–50

Woolley, H., Dunn, J., Spencer, C., Short, T. and Rowley, G. (1999a) 'Children describe their experiences of the city centre a qualitative study of the fears and concerns which may limit their full potential', *Landscape Research*, vol 24, no 3, pp287–301

Woolley, H., Dunn, J., Spencer, C., Short, T. and Rowley, G. (1999b) 'The child as citizen: Experiences of British town and city centres', *Journal of Urban Design*, vol 4, no 3, pp255–282

Yabe, N. (2003) 'Population recovery in inner Tokyo in the late 1990s: A questionnaire survey in Minato Ward', *Human Geography*, vol 55, no 3, pp79–94

7

Service Space

> One should be able to play everywhere, easily, loosely, and not forced into a 'playground' or a 'park'... The failure of an urban environment can be measured in direct proportion to the number of playgrounds. (Ward, 1978, p73)

Service Space for Children: Not Just the Park and the Playground

Reflecting on changes in their suburbs since they were children, a group of relatives of one of the authors realized that many of the places they remembered in their cities during childhood were no longer there. The corner stores that seemed to be on every second block were mostly gone. Blocks of residential units had replaced smaller shopping centres and hardware stores. Many of the local hotels, where locals (mainly men) had congregated after work to share a beer or two, had been demolished. Some churches had disappeared, too, along with post offices, doctors' surgeries, health clinics, tennis courts and even some swimming pools. For most of these services, smaller ones in the local neighbourhood had been replaced by larger facilities that were more centrally located. The children's playgrounds were mainly still there, but with less and different play equipment, and often with fences. How did these changes affect children's worlds in modern cities? This chapter explores the role of service spaces for children.

To children, cities represent an invitation to explore a myriad of new experiences in a range of service spaces (Kalev, 2008). Several spaces can be defined as service spaces for children: they provide specific services for children, either for play, health, entertainment or education (other than through schools). These spaces include playgrounds, health centres, shopping centres, libraries, galleries and museums, swimming pools and community gardens. As well as these spaces that provide particular services, we should also add other places that provide opportunities for children, including sidewalks or footpaths, which children use not only as a corridor for movement, but as a site for play and social interaction. Children can and do use any space as a play space, despite the efforts of adults to contain their activities to certain spaces or to control their behaviour in other spaces. Indeed, as Colin Ward (1978) pointed out in *The Child in the City*, many spaces that are not even noticed by adults are favourite spaces for children's play and socializing. We could also add schools to this list, not as places for education, but as playground spaces in the out-of-school hours (the recent trend towards constructing high fences, sometimes topped with barbed wire or spikes, reduces the possibility of using schools as such a service space; see Figure 7.1).

Children can and do use any space as a play space, despite the efforts of adults

Figure 7.1 A Sydney school with a high fence, topped with barbed wire: Note the wire pointing inward towards the children

Source: Paul Tranter

The way in which these service spaces are accessed, designed and controlled (usually by adults) can either enhance or frustrate children's experiences of these spaces. Some of these spaces are used by many children on a daily basis, while others are used only rarely (e.g. health services).

The more service spaces that there are in a neighbourhood, the greater the chance that children (and adults) will make contact with other children (and adults); hence, the greater the level of communality or social capital. Some researchers have identified service spaces as 'third spaces' (Levy, 2008), between home and school, which are an important intermediate sphere between private and public spaces. These third spaces provide non-formal and non-planned settings for children to participate in the social life of the community. One important function of these spaces is to allow children to 'hang out' with each other. 'Hanging out with friends' is often identified as a favoured pastime of children, and children are frustrated if service spaces that allow this are not available (Corcoran et al, 2009).

The way in which we conceptualize children influences how these spaces are designed, located or operated. For example, if we see children (or young people) as a nuisance or as deviants, then their service space is likely to be located in spaces out of the way of adults. The location of skate parks can provide clues as to how we conceptualize children and youth: they are often built out of the way, in leftover or unwanted land, behind shopping centres or sports grounds.

Many of the service spaces for children are provided by the state (e.g. healthcare facilities, playgrounds, museums). Due to the dominance of neo-liberal policy agendas in many nations (particularly the UK, the US, Australia and New Zealand) (Gleeson and Low, 2000), many of these services are being closed, amalgamated or are not being built, particularly in new suburban developments. Thus, these service spaces are often enduring the same fate as local schools. The dominant planning model in these countries has moved away from a social democratic model to one dominated by market interests. Such interests are hostile to notions of the public interest. Leaving urban planning to the logic of the market relegates the provision of services (or collective goods) for the community to a low priority. When children are considered under the neoliberal ideology, the main concern is about keeping public space free of trouble-making children (White, 1996; Malone and Hasluck, 2002).

Gleeson (2006) sees neoliberalism as being responsible for an erosion of the quality of the public sphere that services children's needs. It is typified by 'heightened individualism [and] fading respect for the public realm' (Gleeson, 2006, p36). This is a particularly worrying issue for children in low-income areas where young people rely more heavily on public space for social interaction due to limited resources in their families to access commercialized spaces.

Another factor affecting the provision of children's service space is the increasing densification of the city. In some cases this can increase the provision of local services, particularly if they are dependent upon a critical mass of people within a particular spatial range. However, the pressure to increase residential densities can also lead to the loss of local open spaces, and children are likely to be acutely affected by this.

Although the state is the main provider of service space for children, there is an increasing market for commodified play experiences for children (McKendrick et al, 2000). Another way of looking at this commercialization of play experiences, as we argue below, is that a market exists in segregating children from adults. Thus, commercialized play spaces are being constructed, often within major shopping centres or sporting or recreational centres. Parents use these to 'drop children' so that they can go shopping, go to the gymnasium or have a coffee. Yet, it is not only the commercialized play spaces that have been critiqued by researchers. The whole concept of 'playgrounds' has been questioned.

Playgrounds

Playgrounds are the most obvious example of a service space designed specifically for children, though adults can also have fun in them (see Figure 7.2). The planning of urban areas in most developed nations includes provision for children's playgrounds. Town planners carefully consider the distribution of playgrounds of various sizes throughout the city. Typically, each suburb has a number of small pocket playgrounds, usually with a few pieces of fixed equipment. Then there are a smaller number of larger better-equipped playgrounds serving the needs of a larger region of the city. There is typically a large city park, which often combines formal gardens with children's play equipment. Playgrounds have been the subject of extensive research, including research that critiques their role in the lives of children (Cunningham and Jones, 1999; Maxey, 1999). Such research argues that rather than providing valuable play experiences for children, many playgrounds are bland and boring, and can contribute to a process referred to as 'childhood ghettoization' (Matthews, 1995, p457). Playgrounds in this sense are seen not as service spaces for children as much as places that provide a service for adults – keeping children out of the way of the productive work of adults.

Playgrounds can contribute to 'childhood ghettoization'

Figure 7.2 Adults as well as children can enjoy conventional playgrounds (Amsterdam, The Netherlands)

Source: Claire Freeman

As Roger Hart explains: 'the only time playgrounds in the USA are really exciting for children seems to be when they are being built, for there are lots of materials for them to work with' (Hart, 1995, cited in Maxey, 1999, p20). Maxey (1999, p20) goes on to argue that playgrounds 'help produce and maintain divisions between adult and child'. As implied in the opening quote for this chapter by Ward (1978), if a city needs 'playgrounds', adults are effectively saying to children: 'Sorry, kids, but this is our city. Oh, we have left these little bits of it for you. Look, there's a playground there with a little blue swing. But if you want to get to this playground, you've got to come to see us, because we've made the rest of the city too dangerous for you.' Colin Ward's quote suggests that the whole city should be the play space for children, and if it is not, then we should be rethinking the way in which we use city spaces to make them safer for children. As Cunningham and Jones (1999) argue, the playground is a confession of the failure of city planning.

Some researchers have even argued for the 'un-planning' of play provision on the grounds that children will create their own play environments (Ward, 1978). Yet, despite the appeal of the arguments about the inadequacies of playgrounds in meeting the needs of children, there is one area where there is arguably a need for careful planning: meeting the play needs for children of *all* abilities.

Playgrounds for Children of *All* Abilities

A recent focus of research and policy is the design of playgrounds and play spaces for children of *all* abilities (Department for Victorian Communities, 2007; UNICEF, 2007). Children with disabilities, as well as their families, 'constantly experience barriers to the enjoyment of their basic human rights and to their inclusion in society' (UNICEF, 2007, piv). For children, one area where this is particularly evident is in playgrounds. While playgrounds may not have been designed with an understanding of children's needs for play (and creativity), most playgrounds provide even less opportunity for children with disabilities. Hope that this might change is found in evidence of an increasing international awareness of children's rights, as well as the rights of persons with disabilities. Two major international treaties are an important part of this increasing awareness: the United Nations Convention on the Rights of the Child, and the Convention on the Rights of Persons with Disabilities. The first of these sets out the basic rights of children and the obligations of governments to fulfil those rights. All except two countries have ratified this treaty – Somalia and the US. Two articles of the convention are relevant here:

All except two countries have ratified the UNCROC – Somalia and the US

1 Article 2: 'State Parties shall respect and ensure the rights set forth in the Convention to each child ... irrespective of the child's ... disability.'
2 Article 23: 'ensure that the disabled child has effective access to and receives education, training, health care services, rehabilitation services, preparation for employment and recreation opportunities in a manner conducive to the child's achieving the fullest possible social integration and individual development.'

The United Nations Convention on the Rights of Persons with Disabilities came into force in May 2008. As of 13 December 2009, it had 144 signatories and 79 ratifications. Australia ratified in July 2008, New Zealand in September 2008 and the UK in June 2009 (United Nations, 2010).

Despite these conventions, children with disabilities are still confronted with daily obstacles to enjoying their rights:

> Even in playgrounds with equipment designed to accommodate their special needs, many of these children cannot navigate the structures without an adult to help them sit supported ... With all these challenges, consequently ... the families, including those with typically developing siblings, simply stay at home. The old term shut-in still applies. (Bergman, 2007, p1104)

A new way of thinking about children and disabilities is now being applied to playground design (Moore et al, 1997). There is now a shift from focusing on the limitations of children with disabilities, to the goal of identifying barriers within society, to inclusion for these children (UNICEF, 2007). When applying this to playgrounds, it is important to recognize that play takes many forms. 'While some children may not be able to swing from monkey bars, they will be able to be included in imaginative play, role play or creative play' (Department for Victorian Communities, 2007, p23). The more nature and natural materials that there are, particularly loose materials, and the greater the ways in which the space can be manipulated by children, the more likely children of all abilities are to find a stimulating play environment. As well as these considerations, fine detail considerations are important, such as wheelchair-accessible benches (without seats), huts/dens and play benches (see Box 7.1).

Box 7.1 Boundless Playgrounds

Several initiatives around the world promote the design of playgrounds to suit children of all abilities. One notable US example is the Able to Play Project, a programme that builds barrier-free playgrounds (known as Boundless Playgrounds), originally throughout Michigan, funded by the Kellogg Foundation. There are now more than 180 Boundless Playgrounds in 31 states and two Canadian provinces.

Boundless Playgrounds aim to create play environments that encourage children with and without disabilities to play together. These playgrounds require at least 70 per cent of play activities to serve children with physical disabilities, allowing for greater 'integration' of all children. Such playgrounds are not just about wheelchair access. They also address the needs of children with sensory and developmental disabilities. They are designed to be fun, rigorous and challenging places for all children – not just special needs kids:

> The process of creating Boundless Playgrounds is also about educating and changing communities ... the more communities work with the people and children with disabilities, and the more young children of all abilities play together, the more all kinds of barriers disappear ... playgrounds become a focal point for both children's play and community change. (W. K. Kellogg Foundation, 2008)

As well as playground design, another issue for playgrounds for children with disabilities is the choice of investment by city governments in either one or two large well-equipped playgrounds, or a number of smaller playgrounds that are more accessible to families. The value of access to local play spaces is recognized by families with children with disabilities. Local spaces can make a real difference to everyday lives: places to meet local neighbours and generate friendships through repeated interactions. However, larger parks and playgrounds are also valued by families with special needs. One innovation for children is the Liberty Swing, a large swing designed so that children confined to wheelchairs can still experience the sensation of swinging (see Figure 7.3). A criticism of the Liberty Swing, particularly as it has been installed (surrounded by large metal fences with locked gates), is that children using the swing are physically and socially segregated from other children, who may only be playing the role of spectators. This makes any attempt at inclusive play very difficult to achieve.

Adventure Playgrounds

Not all playgrounds are bland and boring, or have an emphasis on being 'as safe as possible'. Adventure playgrounds, in their various guises, were first implemented in European inner cities after World War II (Kozlovsky, 2007). The concept of adventure playgrounds was pioneered by a Danish landscape architect (Carl Theodor Sørensen), whose observations of children at play led to the conclusion that children tend to play everywhere except adult-constructed playgrounds. Sørensen saw adventure playgrounds as places where children had a sense of ownership and control over both the play equipment and the play behaviour.

Figure 7.3 A Liberty Swing for use by children in wheelchairs, separated from other park users by a high fence

Source: Paul Tranter

Adventure playgrounds were sometimes referred to as 'junk playgrounds' (Kozlovsky, 2007), where children used whatever loose materials were available for play. Modern adventure playgrounds rely on donations of play resources. However, because play materials do not need to be expensive, the adventure playground model can work effectively in low-income areas.

One very successful adventure playground is known as The Venture, in a deprived estate in Wrexham, Wales (Brown et al, 2007). Established in 1978, The Venture now provides a model for adventure playgrounds throughout the UK and Europe. People from all over the world have visited The Venture and have been inspired to try the same approach. Like all adventure playgrounds, whether in Europe, Japan or the US, The Venture provides opportunities for children to create their own play environment, and provides a space where they can operate on their own terms (Brown et al, 2007). These spaces allow children to build dens (cubbies or huts) and creative structures, and sometimes care for animals. Occasionally adults criticize these playgrounds as being too untidy or dangerous. This may simply be due to adults' lack of appreciation of the importance of real play spaces for children and an obsession with safety. Perhaps they should be working towards making play 'as safe as necessary, not as safe as possible' (Staempfli, 2009).

There are now over 1000 such playgrounds in Europe, mainly in Denmark, Switzerland, France, Germany, The Netherlands and the UK. There are more than 80 adventure playgrounds in London (Staempfli, 2009). Adventure playgrounds in North America tend to have more ready-made play structures in more controlled environments. In North America, the community garden model for adventure playgrounds is also common.

Community Gardens

Community gardens may provide some of the best service spaces for children in terms of giving children a sense of place, particularly, but not only, in disadvantaged urban areas (Peters and Kirby, 2008; Goltsman et al, 2009). The critical factor seems to be whether or not children are made welcome in these spaces, which are sometimes seen as 'adult spaces'. Community gardens have a long history, particularly in the US and Australia (Thompson et al, 2007). They can provide benefits for children's health through increased contact with nature, access to more nutritious foods, and increased social connection with other members of the local community (Allen et al, 2008).

Research on community gardens in upstate New York found that many community members valued local community gardens for their children, and children also had positive views about them: 'Children (in a housing project) see it [the community garden] as an actual piece of land that they have control over, they have pride of ownership' (Armstrong, 2000, p324). In several community gardens in public housing projects, children participate in the gardening and the produce is shared around the children and their families. Sometimes the gardens are seen specifically as children's spaces (while others are seen as retirees' space or space for mothers on welfare). In a survey of the reasons for participating in the community gardens, more than two-thirds of respondents agreed that they provided a good family/children's activity (Armstrong, 2000).

In an earlier study of community gardens in Loisaida, a part of the Lower East Side of Manhattan (Schmelzkopf, 1995), several of the gardens were child or family oriented, with most of these gardens run by women. Even though most of the land was devoted to intensive gardening, there were spaces specifically set aside for the use of children. In this area, known for high crime rates, some gardeners commented: 'if they or their children were not in the gardens, they would be out getting high' (Schmelzkopf, 1995, p373). In this area the garden provided a service space with security, where children could feel safe yet still be outside with other people, and 'the only site in Loisaida where they can be around nature, where they can watch and make things grow' (Schmelzkopf, 1995, p373).

Community gardens can break down the segregation of children and adults

One of the key benefits of community gardens as a service space for children is that they can encourage intergenerational interactions that break down the segregation of children and adults. In a study of community gardens in public housing estates in Sydney, children represented an important proportion of the actual gardeners (Thompson et al, 2007). There were also interesting gender differences in the way in which that adults welcomed or discouraged the involvement of children. Men were more likely to discourage children and young people, while women usually encouraged the children to be involved, and to take part in the gardening process.

An interesting combination of gardens and playgrounds comes with the concept of permaculture playgrounds (Bulut and Yilmaz, 2008). These playgrounds recognize the limitations of traditional playgrounds (outlined earlier in this chapter), and seek instead to provide children with more stimulating and malleable play spaces, where they can construct 'natural play areas using vegetation, animals, topography, water and other natural landscape elements' (Bulut and Yilmaz, 2008, p35). Permaculture, a concept developed during the 1970s by two Australian scientists, is a design system that can be applied to land use, food production or society. It targets sustainability and productivity by integrating human life and ecology. Applying this to children's playgrounds means providing carefully chosen plants for children to interact with. While this may not involve children in gardening as in the community gardens, it provides many of the benefits of children interacting with natural environments discussed in Chapters 4 and 9.

Shopping Centres, Malls and Markets

The number of spaces in cities that may be deemed unequivocally 'public spaces' are dwindling, and this can have implications for children (Lownsbrough and Beunderman, 2007, p14). As private property ownership and the privatization of shopping areas increase, children and young people are increasingly excluded, particularly from larger enclosed shopping malls. This exclusion can be achieved in various ways: curfews, move-on powers, admission charges, camera surveillance and repeated questioning of 'particular types of people', including groups of young people (Crane, 2000). This situation has arisen as a result of children and youth engaging in an activity that has important benefits for them – to 'hang out':

Children are increasingly excluded from enclosed shopping malls

> They often 'hang out' and socialise in the mall, using it as the gathering-place that malls promote themselves as providing. Their presence in malls is obvious during weekend evenings, as the mall has become a place where youth can socialise indoors without the direct supervision of parents; in places with uncomfortably cold or hot climates, malls are important social spaces. (Staeheli and Mitchell, 2006, p986)

Malls have become so attractive to children and youth that many malls have implemented policies to exclude them or to severely limit their access. In some shopping centres on periphery estates in the UK, unaccompanied children attract particular surveillance. Children observed in the centres during school hours are reported to the police (Flint, 2006). The particular policies of surveillance and access control are based on speculation about the likely behaviour of youth (Flint, 2006). In this case, children and youth are clearly conceptualized as nuisances or demons.

Paradoxically, these policies of surveillance and exclusion may have added to the dominance of larger malls in the lives of children. In a world where 'public space' is increasingly seen as dangerous, particularly for children, malls are seen as 'policed' private space (Corcoran et al, 2009). Both parents and children may feel safer in malls than on public streets in traditional shopping centres, as illustrated in the example below from Elizabeth, a disadvantaged outer Adelaide suburb. In this example, we can see both the conceptualization of children as dangerous, as well as the development of images of fear associated with smaller shopping centres.

In a study of 'mall walking' in Elizabeth, South Australia, researchers tell a story of a widower in his 70s who pointed to a group of school children, explaining that 'he was scared of young people' (Warin et al, 2008, p193). The widower reported a sense of security coming from walking in the mall in a group. This mall had been marketed, as are many malls, as a place of warmth, safety and convenience, implying an escape from the conditions of other (smaller) local shopping centres that evoked an image of danger: 'One small shopping centre is surrounded by 12 foot fences that are locked at night, and others have meshed store windows with shopkeepers serving behind wire screens' (Warin et al, 2008, p193).

In some areas, where the local main street shops have not become areas of fear and danger, children can still hang out together, as in the main streets of some outer suburbs and small villages near Dublin (Corcoran et al, 2009). But even small-town shopping centres and village squares are not immune from the trend towards privatizing public space. Private interests have noticed the popularity of some traditional town squares, particularly in many European cities, and have attempted to create shopping spaces that emulate these squares. For example, Orchard Square in Sheffield resembles a traditional town square, but 'its purpose is exclusively commercial', and the aim is to attract 'consumers' (Lownsbrough and Beunderman, 2007, p14). Truly public spaces are 'civil' spaces as well as public – meaning that they are spaces where people are content to linger, to be idle and to interact with each other or with the features of the landscape (e.g. fountains). Instead, many quasi-public spaces created by the private sector are places that are more transient: they may be more awe-inspiring, but people are busy and active and are there with a purpose – usually to shop (Bauman, 2000). Larger shopping centres may be attractive to many families with children; but they are often inaccessible to children travelling independently when they are surrounded by a sea of car parks and cars.

Truly public spaces are 'civil' spaces where people linger and interact with each other

As well as shopping centres and malls, local markets are experiencing a resurgence in some areas. Local markets are typically located in a low-rent location (often in car parks that are empty on weekends) and recur periodically (e.g. weekly or monthly). Stallholders set up in a similar position each time. The markets develop a familiarity, and stallholders and patrons get to know each other after repeated interactions. Children are an important part of the landscape of this type of service space (see Figure 7.4).

Sophie Watson (2009) argues that markets have been neglected in the research on public space, particularly as a site of social interaction. The social connections in markets offer a contrast to the decline of social contact reported in shopping malls, and also provide more possibility for inclusion of marginalized groups and the co-mingling of different groups, including different age groups.

Watson sets the context for her research on markets in eight locations in the UK by noting the 'pessimistic accounts of the decline of public space as a consequence of the drive towards privatisation or thematisation ... and the prevalence of a risk culture' (Watson, 2009, p1578). She argues that markets can potentially provide an alternative social space to alienating shopping centres in cities. An important part of the social experience offered by most of the markets in Watson's study is performance and theatricality, where a festive, sometimes spontaneous, carnival atmosphere is created, and where the usual conventions of retailing can be suspended. Children can be involved in this atmosphere as much as adults, and there are few limitations on children's behaviour in markets compared to shopping malls (where even the theatricality

Figure 7.4 Farmers' markets provide excellent places for children to connect with a very active public space

Source: Claire Freeman

is staged and planned). In Watson's research, parents saw the markets as a safe environment for their children, and sometimes stall owners looked after children while their parents shopped across the walkway. Modern markets in urban areas may resemble markets in a medieval town. In these markets, there are few areas that children cannot use as a play space. In contrast, modern shopping centres identify particular spaces for children – 'playgrounds' inside the shopping complexes.

Commercialized Play Centres, and Play Centres within Large Shopping Centres

Play spaces for children are usually associated with the spaces of the home, the school and the neighbourhood. Homes have backyards to play in and increasingly have 'playrooms'. Schools have designated playgrounds, and local suburbs typically have a range of playgrounds. Yet, apart from these spaces, children's play and mobility have been increasingly restricted (see Chapter 10). Partially in response to these restrictions, commercial playgrounds for children have sprung up in restaurants, shopping malls and big retail outlets or hardware stores, airports, gymnasiums and fitness centres, and hospitals. These are usually 'add-ons' and can be outdoors or indoors. They are usually segregated from adult spaces, often by high fences or walls or glass, and adults can observe children in these spaces in much the same way as they would observe monkeys in a zoo enclosure. In some cases, 'adult viewing zones' are built alongside the playgrounds.

These commercialized playgrounds can be seen in contrasting ways. One view is that they allow children to accompany their parents to places that otherwise parents would feel unable to take their children to (because of social conventions about children being out of place in many of these spaces). An alternative view is that they continue the practice of segregating children from adults in the same way as discussed above for traditional playgrounds. Another view is that these commercialized playgrounds are not for children at all. They are primarily designed as a way to attract more customers to a commercial space, and they perpetuate the view that special spaces should be provided for children (McKendrick et al, 2000). This raises questions about whose interests are being served by these commercial playgrounds.

Most of the commercialized playgrounds cater for younger children. Another form of commercial child space involves video game arcades (also called amusement arcades or, in Japan, game centres). Game centres in Japan are still very popular entertainment spaces for children. They not only attract children, but also businessmen and women, and housewives (Eickhorst, 2006). Although children have been lured away from these game centres by the recent growth in home computer games (in Japan and in nations throughout the world), these spaces continue to survive, and they provide a space where there is little segregation between adults and children (or between different socio-economic levels). They differ from the commercial playgrounds in that they provide a service space where children of all ages can congregate and socially interact, albeit with the distraction of the virtual world of the video games. Another type of space that can involve play concerns healthcare services.

Hospitals and Healthcare Services

Healthcare services (ranging from local health clinics and general practice doctors' surgeries, to large children's hospitals) provide an important service space for children in multiple ways. Not only do they provide healthcare, but they also provide places for social interaction and places for play. Play has long been recognized as an important part of the recuperative process for children in hospitals (Harvey, 1975), and play in natural spaces has been found to enhance recovery of children in hospital (Maller et al, 2002; Schweitzer et al, 2004). Even in hospitals, play provision is now seen as a commercial opportunity.

Play is an important part of the recuperative process for children in hospitals

There has been growing recognition in policy of the importance of children's needs in the design and operation of hospitals (Adams et al, 2010). As well as having specialized children's hospitals in cities, all hospitals in the UK are now required to provide 'child-friendly' services (Kmietowicz, 2003). These include dedicated children's units in emergency departments and play areas for children on wards. Given the particular needs and challenges of patients and staff in hospitals, it is difficult to argue that the whole hospital should be a play space. Thus, play areas for children (preferably soundproofed) are a sensible idea for hospital service space. Yet, play can be seen in various ways, and it is important that spaces in hospitals have a welcoming atmosphere that helps children to feel at home. In order to achieve this, there have been some notable success stories involving children in the design of more child-friendly hospital environments (see Box 7.2).

Another issue on a larger spatial scale than the emergency rooms discussed in Box 7.2 is the access that children have to healthcare services in their local

Box 7.2 A more child-friendly hospital: Input from Sullivan Elementary School, Massachusetts, US

Few children see hospitals as inviting places to visit (Adams et al, 2010). As part of a broader project to involve local school children with their communities, a community-based learning experience in a school in Massachusetts, US, engaged children in making the local hospital more child friendly. In a cost-saving restructure, the paediatric section of the North Adams Regional Hospital had been closed, and the emergency department became the primary healthcare facility for many children. A starting point for the project was that 'children expressed fear and discomfort around their experiences in the hospital' (Bartsch, 2001, p3). Deborah Coyne and Roberta Sullivan from Sullivan Elementary School designed a project to involve kindergarten children in the process of making the hospital more child friendly.

While the main objective of the project was to enhance children's learning experiences (e.g. cooperative learning skills), an important outcome of the project was a more child-oriented hospital, albeit without its paediatric ward. The aim was to create a less threatening and more nurturing hospital space for a child to be with their family.

The school gained approval and support from the hospital to create a more welcoming emergency room for children, and a partnership was established between Sullivan School and the North Adams Regional Hospital Day Care. The project began with a tour of the hospital with their teacher, and a nurse explained to the children the hospital safety precautions. The children then interviewed hospital personnel, from the chief executive officer (CEO) to housekeeping staff. This contact with staff reduced the children's own anxieties about hospital emergency rooms. As the project developed, the students agreed that children would feel more at ease in an emergency room with familiar works of art and books. So they decided to change the appearance of a waiting room by using their own artwork:

> The hospital emergency room served as a place of celebration for a special group of kindergartners determined to 'cheer up' the surroundings. With an air of maturity, these children climbed ladders to secure their artwork. They explained to the nurses, doctors, janitorial staff, and cafeteria workers why they were there and what made their work worthwhile. (Bartsch, 2001, p3)

Doctors and nurses, who credited the students with reductions in the need to use a restraining board with uncooperative children, saw the children's project as very successful. In an encouraging follow-up to the project, several other community agencies requested student artwork for their facilities. These service spaces include day-care centres, the police station and courtroom waiting rooms, and doctors' offices.

area. In the same way that larger shopping centres are taking over from smaller ones, larger (and less personal) hospitals are taking over from smaller health centres. Smaller suburban medical practices are also under increased competition from large medical centres. This can be understood in terms of the dominance of a bio-medical model of health over a socio-ecological one (Kearns, 1993).

In the bio-medical model of health and healthcare, patients are seen in much the same way as mechanics would see cars. If someone is sick, they are taken to a doctor or hospital so that medical intervention can solve 'the problem'. In this way, medical care is a curative approach and is based on linear thinking. In contrast, the socio-ecological model is based on a preventative approach to health and healthcare. Instead of linear thinking, holistic thinking is used, and health is seen as the result of the interrelationship between people and their environment. The environment here includes the physical environment, as well as the social, economic and cultural environment.

Arguably, the socio-ecological model is more appropriate for a child-friendly perspective, as it recognizes the health benefits of allowing children the freedom to playfully explore their environment, rather than simply relying on medical intervention to help when something goes wrong with the child (the bio-medical view). In the socio-ecological model, while healthcare facilities are seen as important, so, too, is the range of other service spaces – even museums, art galleries and libraries.

Museums, Art Galleries and Libraries

Museums, art galleries and libraries provide important service spaces for children in terms of their role in enhancing education and learning. The potential for using these spaces for learning outside the classroom is vast, and education researchers encourage an increasing use of these spaces (Malone, 2008). As well as being spaces for learning, museums, galleries and libraries are increasingly seen as places for play (see Figure 7.5).

Directors of museums, art galleries and even libraries have for several years been looking for ways to make these service spaces more appealing and more relevant to children (Appleton et al, 2009). An important part of this search is the conceptualization of children. Rather than seeing children as blank slates to be educated, or as nuisances to adults enjoying the culture and ambience of these service spaces, children are now more likely to be regarded as individual playful

Figure 7.5 Children being planes in the Science Museum in London, UK

Source: Robin Quigg

beings (with widely varying interests) and as capable social actors. *The Guardian* newspaper in the UK published a 'Kids in Museums Manifesto', an article listing 20 suggestions for making museums more child friendly (*The Guardian*, 2003) (see Box 7.3). This list reflects this changing conceptualization of children. These suggestions have relevance to art galleries and libraries as well.

Examining the list in Box 7.3, we can identify several clues to good practice about providing service space for children, some of which apply specifically to children and others to all visitors. The concept of children as capable social actors is recognized in point 5 (consult with children) and point 20 (don't make presumptions about what children do and don't like), which also recognizes that 'children' represent a divergent range of interests and abilities, and should not be seen as a homogeneous group. The list also indicates the importance of not segregating children and other visitors, and having a range of types of spaces, including space for active play and for quiet reflection. Thus, many of the principles discussed in Chapter 4 regarding good school ground environments can also be applied to these service spaces. Perhaps most importantly, the most child-friendly museums are those that encourage children to be children.

Some countries have made particular efforts to ensure that their museums and art galleries are child friendly. In the UK, the Kids in Museums organization promotes child-friendly visits to museums and art galleries across the country.

Box 7.3 'Kids in Museums Manifesto'

1. Be welcoming.
2. Be interactive and hands on.
3. Be pushchair accessible.
4. Give a hand to parents.
5. Consult with children.
6. Be height aware.
7. Have lots of different things to do.
8. Produce guides aimed at children.
9. Provide proper good-value food, highchairs and unlimited tap water.
10. Provide dedicated baby changing and breast-feeding facilities.
11. Teach respect: help children to learn that there are objects they should not touch.
12. Sell items in the shops that are not too expensive and not just junk, but things that children will want to treasure.
13. Have flexible family tickets.
14. Provide some open space – inside and outside – where children can run about and let off steam.
15. Provide some quiet space where children can reflect.
16. Make it clear to child-free visitors that the museum is family friendly.
17. Have dedicated family-friendly days when extra activities are laid on for kids, and those who want to avoid the crowds can choose not to attend.
18. Provide a crèche for young children at major museums.
19. Attract all ages, from toddlers to teenagers, without offering separate facilities for each.
20. Don't make presumptions about what children do and don't like. Some kids can appreciate fine art as well as finger painting.

Source: adapted from The *Guardian* (2003)

Kids in Museums was established after the son of Dea Birkett (a British writer) was ejected from an exhibition of the Aztecs at the Royal Academy because he shouted 'Monster!' at an exhibit. This organization has compiled the 'Kids in Museums Manifesto' (see Box 7.3), has its own website and organizes awards for the UK's most family-friendly museums.

For museums, it is not only the qualities of the space that are important to children. While children can and do benefit from policies to make them more usable for children, access to these service spaces is also critical. Ideally, they should be located where children can independently access them, particularly via safe and reliable public transport (see Chapter 10). While there has been considerable recent research and policy debate about children's use of museums and galleries, there has been little research on another type of service space – prisons (see Box 7.4).

Box 7.4 Prisons as a children's service space

It may appear incongruent for prisons to be discussed as children's service space. Yet, prisons provide a very real service space for a large (and growing) number of children: access to their incarcerated parents. When this issue is addressed in policy, it usually concerns the access of children to their mothers. This tends to exclude fathers from consideration, and the majority of prisoners are male (92 per cent in the US) (Glaze and Maruschak, 2008). In the US, two-thirds of fathers in prison have never received a visit from their child, while only 5 per cent thought that the child did not want to visit (Rosenberg, 2009).

Steady increases in imprisonment rates are found in most Western industrialized countries over the last few decades. Yet, the contact that prisoners have with their children has decreased. An estimated 700,000 children are separated from parents in prison in the European Union. In the US, the figure is over 800,000, and in the UK over 100,000. Some estimates for the US put the figure at over 1.5 million (Seymour and Hairston, 2001).

Visits by children to their parents in prison affect both imprisoned mothers and fathers. Evidence indicates that it is harder for children to keep contact with fathers than mothers. Access by children to their parents in prison is limited by distance, unavailability of public transport, financial difficulties, lack of cooperation from departments of corrections, inflexible visitation hours, and the lack of child-friendly visiting conditions – particularly 'visiting procedures that are uncomfortable or humiliating' (Parke and Clarke-Stewart, 2001; Seymour and Hairston, 2001, p13).

There is evidence that children do want more contact with their parents, and there are some schemes in Europe and the US that have tried to enhance the contact of prisoners with their families. Observers of the notorious men's prison in South London, Wandsworth, will notice, on school afternoons, groups of 13- to 15-year-old children walking into the prison unescorted by adults. These children are taking part in 'homework visits', where children sit down with their fathers and do homework. The visit is structured around an activity that is officially sanctioned, and allows the fathers to get to know what their children are doing at school, enabling the sharing of knowledge (e.g. numeracy skills) between fathers and children. The fathers in prison thus become functional members of the school community.

Research in the UK suggests that while children were positive about visiting their fathers in prison, they were not always as positive about the visiting arrangements. Access to incarcerated fathers, despite the unique context of prisons, can have helpful impacts upon children's development (Dyer, 2005).

Children's Views on What Makes a Good Museum

Within the topic of service space for children, children in schools in Canberra were asked about the things that would make the following places good places for children: playgrounds, sports ovals and museums. The responses to museums were the most intriguing, see Table 7.2. For children, the things that make a museum a good place for children were that it should be interactive, safe and fun (not boring). There should be different displays for different ages, and displays should be designed so that children can enter them and get a feel for what the places and things are really like. Figure 7.6 depicts a sign with the words 'Please don't touch', with the 'don't' crossed out. In this drawing, children are touching, drawing and entering the exhibits, and there is a special place for younger children at the museum.

Table 7.2 Children's views on what makes a good museum

Sculptures; interactive or hands on; help make art.	A place where you can drop your kids off and they will be safe; fun stuff for kids, not just adults; interactive touch-screen so that kids can understand; a gaming museum like PS3, S2, PSP, Xbox.
Interactive activities and games.	There should be a virtual gaming area for creating your own game.
It's fun and interactive.	I wish for more things that you can touch, and more kid friendly.
Artefacts teach us history.	A safe place where we could keep kids; more child friendly; less complicated information and a place we can eat.
Kids can make things for the museum.	Using machines that people used a long time ago.
History is interactive.	I want more stuff you're allowed to touch.
Hands-on stuff instead of boring writing; lots of security because it makes me feel safe.	In a museum, I think there should be an area just for kids where things are simply explained and there are hands-on activities to do with that piece where kids can apply what they have learned from that particular item, while their parents browse outside and around the museum.
I think a museum should have more life-like experience – for example, if it was a space exhibit, then there should be a realistic space ship which you could go inside of, to see what it's really like.	Things that make you feel like you're really there; displays and not just pictures; quizzes on what you've learned; maps; you have to follow a path and find secret compartments that have little displays in them.
I would like more hands-on activities for us; have some food shops.	
More hands-on activities like touchable guns and planes and short moving pictures.	
Museum with a variety of displays; dinosaurs; peaceful.	
More stuff to do in the museum.	
Kids should have their own guide that will go with them to the place they want to go.	

The Future of the Suburbs: Erosion of Service Space?

The service spaces discussed in this chapter are important in the lives of children for their health, education and play. The presence of children in these places also changes the nature of the spaces, sometimes in ways that are not concordant

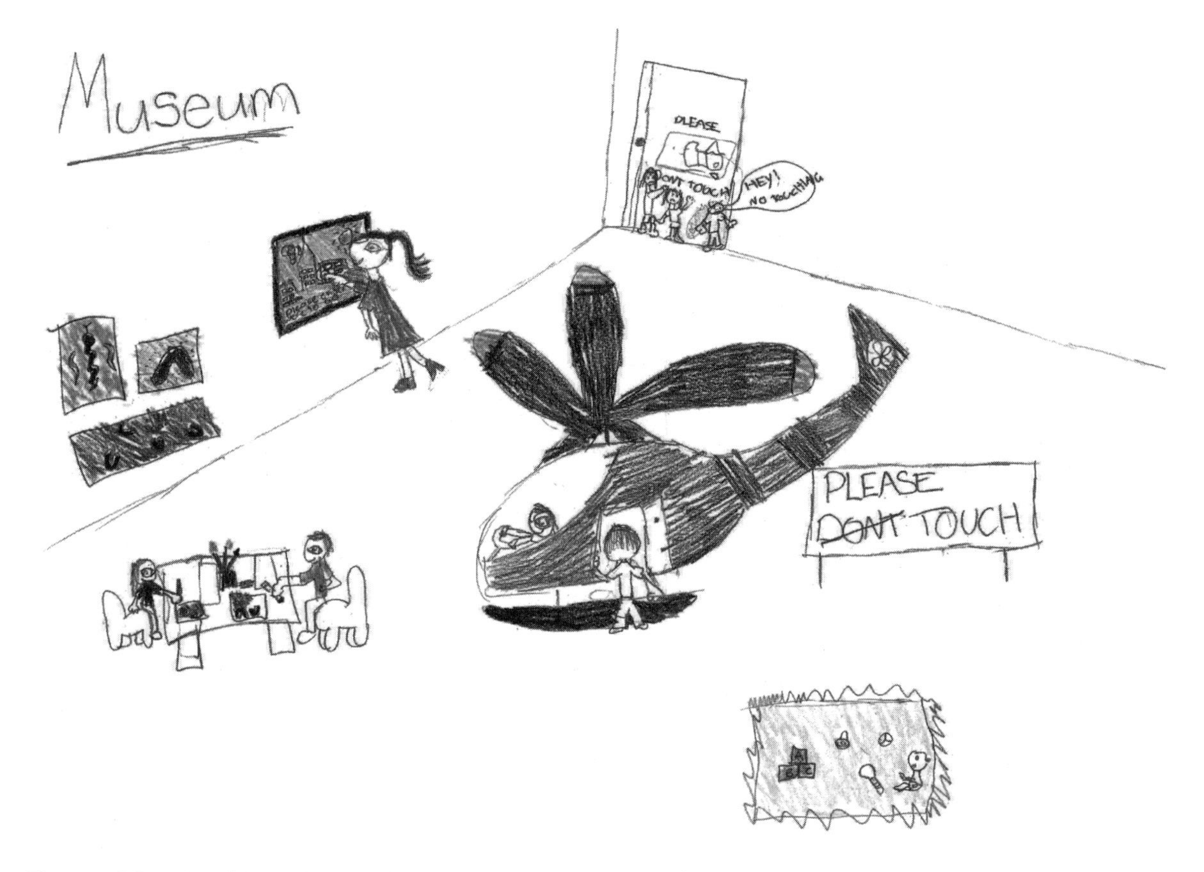

Figure 7.6 Drawing of things that would make a museum a good place for children

Source: Year 5/6 child at Ngunnawal Primary School, Canberra

with the views of adults who operate the spaces or profit from them. In some cases, inspiring models of how to provide for children's play exist (e.g. The Venture adventure playground in Wales). However, for many other service spaces (e.g. shopping centres), the last few decades have involved an erosion of both the number and quality of these spaces in terms of their value for children.

This erosion has arguably been most noticeable in suburban environments, where car-dominated urban forms mean that local service space for children is replaced by more distant services (in much the same way as has happened for services for adults). The dominance of the suburb in many cities is accompanied by the rise of out-of-town shopping centres, business parks, large medical centres and leisure complexes: 'Children's lives become a fragmented mosaic of places' (Freeman, 2006, p73). Yet, not all cities or neighbourhoods have succumbed to these trends. Many cities in Europe, Japan and South America, and several neighbourhoods in the US, Australia, Canada and New Zealand, retain a diverse mixture of service spaces within local neighbourhoods. These provide the opportunity for children to independently access the services and opportunities provided by these spaces, instead of being locked into the adult-dependent access required for many suburban children (see Chapter 10).

The American social commentator James Kunstler presents a challenging argument about the way in which suburbs have developed in the US (the argument could also be applied to suburbs in Australia, Canada and New Zealand):

> ... suburbs represent the greatest misallocation of resources in the history of the world ... The project of suburbia represents a set of tragic choices because it is a living arrangement with no future. (Kunstler, 2005)

One problem for future suburbia will be children's access to service space

Kunstler argues that the medieval town may be a more appropriate model for the living arrangements of the future. There may also be lessons from medieval towns for a model of service space for children. One of the problems for suburbia in the future will be children's access to service space. In the medieval town, not only were children constructed very differently from today's children, they had access to most of the spaces in the city: the whole city was a service space for children.

References

Adams, A., Theodore, D., Goldenberg, E., McLaren, C. and McKeever, P. (2010) 'Kids in the atrium: Comparing architectural intentions and children's experiences in a pediatric hospital lobby', *Social Science & Medicine*, vol 70, no 5, pp658–667

Allen, J., Alaimo, K., Elam, D. and Perry, E. (2008) 'Growing vegetables and values: Benefits of neighborhood-based community gardens for youth development and nutrition', *Journal of Hunger & Environmental Nutrition*, vol 3, pp418–439

Appleton, P., Bezzina, H., Hodge, K., Pollnitz, E. and Stewart, L. (2009) *Kids Matter: Safe and Secure Cultural Environments for Children in the Australian Capital Territory, CMDP Report*, National Library of Australia Staff Papers, Australia

Armstrong, D. (2000) 'A survey of community gardens in upstate New York: Implications for health promotion and community development', *Health and Place*, vol 6, no 4, pp319–327

Bartsch, J. (2001) 'Emergency room', in *Community Lessons: Promising Curriculum Practices*, Massachusetts Department of Education, Malden, MA, pp1–11

Bauman, Z. (2000) *Liquid Modernity*, Polity, Cambridge

Bergman, A. (2007) 'The right of children with disabilities to have fun', *Archives of Pediatrics & Adolescent Medicine*, vol 161, no 11, p1104

Brown, F., King, M. and Tawil, B. (2007) *The Venture: A Case Study of an Adventure Playground*, Pearson Education, Harlow, UK

Bulut, Z. and Yilmaz, S. (2008) 'Permaculture playgrounds as a new design approach for sustainable society', *International Journal of Natural and Engineering Sciences*, vol 1, no 2, pp35–40

Corcoran, M., Peillon, M. and Gray, J. (2009) 'Making space for sociability: How children animate the public realm in suburbia', *Nature and Culture*, vol 4, no 1, pp35–56

Crane, P. R. (2000) 'Young people and public space: Developing inclusive policy and practice', *Scottish Youth Issues Journal*, vol 1, pp105–124

Cunningham, C. and Jones, M. A. (1999) 'The playground: A confession of failure?', *Built Environment*, vol 25, no 1, pp11–17

Department for Victorian Communities (2007) *The Good Play Space Guide: 'I Can Play Too'*, Sport and Recreation Victoria, Melbourne

Dyer, W. (2005) 'Prison, fathers, and identity: A theory of how incarceration affects men's paternal identity', *Fathering: A Journal of Theory, Research, and Practice about Men as Fathers*, vol 3, no 3, pp201–219

Eickhorst, E., (2006) *Game Centers: A Historical and Cultural Analysis of Japan's Video Amusement Establishments*, MA thesis, University of Kansas, Kansas, US

Flint, J. (2006) 'Surveillance and exclusion practices in the governance of access to shopping centres on periphery estates in the UK', *Surveillance & Society*, vol 4, no 1/2, pp52–68

Freeman, C. (2006) 'Colliding worlds: Planning with children and young people for better cities', in B. Gleeson and N. Sipe (eds) *Creating Child Friendly Cities*, Routledge, London

Glaze, L. and Maruschak, L. (2008) *Parents in Prison and Their Minor Children*, Bureau of Justice Statistics Special Report, Publication no NCJ-222984, US Department of Justice, Washington, DC

Gleeson, B. (2006) 'Australia's toxic cities: Modernity's paradox', in B. Geelson and N. Sipe (eds) *Creating Child Friendly Cities*, Routledge, London, pp33–48

Gleeson, B. and Low, N. (2000) *Australian Urban Planning: New Challenges, New Agendas*, Allen and Unwin, Sydney

Goltsman, S., Kelly, L., McKay, S., Algara, P. and Wight, L. (2009) 'Raising free range kids: Creating neighborhood parks that promote environmental stewardship', *Journal of Green Building*, vol 4, no 2, pp90–106

The Guardian (2003) 'The Kids in Museums Manifesto', *The Guardian*, 6 September 2003, www.guardian.co.uk/travel/2003/sep/06/kidsinmuseumscampaign.museums

Harvey, S. (1975) 'Play for children in hospital', *International Journal of Early Childhood*, vol 7, no 2, pp185–187

Kalev, S. (2008) *Child in the City*, Congress of Local and Regional Authorities of the Council of Europe, Strasbourg

Kearns, R. A. (1993) 'Place and health: Toward a reformed medical geography', *The Professional Geographer*, vol 45, no 2, pp139–147

Kmietowicz, Z. (2003) 'Hospitals must be child friendly, says report', *British Medical Journal*, vol 326, no 7394, p840

Kozlovsky, R. (2007) 'Adventure playgrounds and postwar reconstruction', in M. Gutman and N. De Coninck-Smith (eds) *Designing Modern Childhoods: History, Space, and the Material Culture of Children*, Rutgers University Press, New York, NY

Kunstler, J. H. (2005) Speech to the Second Vermont Republic, Vermont

Levy, R. (2008) '"Third spaces" are interesting places: Applying "third space theory" to nursery-aged children's constructions of themselves as readers', *Journal of Early Childhood Literacy*, vol 8, no 1, pp43–66

Lownsbrough, H. and Beunderman, J. (2007) *Equally Spaced? Public Space and Interaction between Diverse Communities: A Report for the Commission for Racial Equality*, Demos, London

Maller, C., Townsend, M., Brown, P. and St Leger, L. (2002) *Healthy Parks, Healthy People: The Health Benefits of Contact with Nature in a Park Context: A Review of Current Literature*, Faculty of Health and Behavioural Sciences, Deakin University, Burwood, Australia

Malone, K. (2008) *Every Experience Matters: An Evidence Based Research Report on the Role of Learning Outside the Classroom for Children's Whole Development from Birth to Eighteen Years*, Report commissioned by Farming and Countryside Education for the UK, Warwickshire

Malone, K. and Hasluck, L. (2002) 'Australian youth: Aliens in a suburban environment', in L. Chawla (ed) *Growing Up in an Urbanizing World*, Earthscan/UNESCO, London, pp81–109

Matthews, H. (1995) 'Living on the edge: Children as outsiders', *Tijdschrift voor Economische en Sociale Geografie [Journal of Economic and Social Geography]*, vol 86, no 5, pp456–466

Maxey, I. (1999) 'Playgrounds: From oppressive spaces to sustainable places?', *Built Environment*, vol 25, no 1, pp18–24

McKendrick, J. H., Bradford, M. G. and Fielder, A. V. (2000) 'Kid customer? Commercialization of playspace and the commodification of childhood', *Childhood*, vol 7, no 3, pp295–314

Moore, R., Goltsman, S. and Iacofano, D. (1997) *Play for All Guidelines: Planning, Design and Management of Outdoor Play Settings for All Children*, MIG Communications, Berkeley, CA

Parke, R. and Clarke-Stewart, K. (2001) 'Effects of parental incarceration on young children', *National Policy Conference*, pp30–31

Peters, E. and Kirby, E. (2008) *Community Gardening*, Brooklyn Botanical Gardens, New York, NY

Rosenberg, J. (2009) *Children Need Dads Too: Children with Fathers in Prison*, Quaker United Nations Office, New York, NY

Schmelzkopf, K. (1995) 'Urban community gardens as contested space', *Geographical Review*, vol 85, no 3, pp364–381

Schweitzer, M., Gilpin, L. and Frampton, S. (2004) 'Healing spaces: Elements of environmental design that make an impact on health', *Journal of Alternative & Complementary Medicine*, vol 10, supplement 1, pp71–83

Seymour, C. and Hairston, C. (2001) *Children with Parents in Prison: Child Welfare Policy, Program & Practice Issues*, Transaction Pub, Piscataway

Staeheli, L. and Mitchell, D. (2006) 'USA's destiny? Regulating space and creating community in American shopping malls', *Urban Studies*, vol 43, no 5, pp977–992

Staempfli, M. (2009) 'Reintroducing adventure into children's outdoor play environments', *Environment and Behavior*, vol 41, no 2, pp268–280

Thompson, S., Corkery, L. and Judd, B. (2007) 'The role of community gardens in sustaining healthy communities' Paper presented to the State of Australian Cities National Conference, Adelaide

UNICEF (United Nations Children's Fund) (2007) *Promoting the Rights of Children with Disabilities*, UNICEF, Florence

United Nations (2010) *Rights and Dignity of Persons with Disabilities*, www.un.org/disabilities/, accessed 17 February 2010

W. K. Kellogg Foundation (2008) *Playgrounds Where All Kids Are Able to Play*, W. K. Kellogg Foundation, Youth and Education Programs, Battle Creek

Ward, C. (1978) *The Child in the City*, Architectural Press, London

Warin, M., Moore, V., Davies, M. and Turner, K. (2008) 'Consuming bodies: Mall walking and the possibilities of consumption', *Health Sociology Review*, vol 17, no 2, pp187–198

Watson, S. (2009) 'The magic of the marketplace: Sociality in a neglected public space', *Urban Studies*, vol 46, no 8, pp1577–1591

White, R. (1996) 'No-go in the fortress city: Young people, inequality and space', *Urban Policy and Research*, vol 14, no 1, pp37–50

8
Cultural Space

> Our children cannot learn respect for diversity, equality and non-discrimination when they are separated from children with disabilities. And our children with disabilities cannot be socialised and learn to become responsible, self determining adults if they do not learn to play amongst the diversity of children in our society. (Bray and Gates, 2000, p39)

Cultural Worlds

Every child is born into a cultural space that can enhance or inhibit their personal development, the family and community they live in and the geographic space they inhabit. Culture can be exhibited in multitudinous ways: the food that is eaten, the language that is spoken, the composition of the family, the games children play, the schools they attend, the festivals they take part in and how and what they celebrate, the place or places they connect to as 'home', and the places they value and take pride in (see Figure 8.1). Culture is hard to define, perhaps because it covers this wide range of life worlds and events. Where culture is defined it is usually with reference to adult rather than child experiences. A definition that has been defined to apply specifically to children is that used by the Queensland government in Australia, and is intended as a guide in the adoption process:

Every child is born into a cultural space that can enhance or inhibit their development

> Culture – describes what people develop to enable them to adapt to their world, such as language, gestures, tools to enable them to survive and prosper, customs and traditions that define values and organise social interactions, religious beliefs and rituals, and dress, art, and music to make symbolic and aesthetic expressions ... Culture determines the practices and beliefs that become associated with an ethnic group and provides its distinctive identity. (State of Queensland, 2009)

Culture has, in some spheres (such as local government literature) often become equated with difference, most commonly race, ethnicity or nationality. However, while culture can and does have these associations, culture is more than cultural difference. It is what characterizes communities, their religious and secular feasts, celebrations, and acknowledgements of life stages such as birth and death. It is how people lead their lives, from how they relate to time, laughter, the natural world, the passing of the seasons, to the clothes they wear and the language they speak. Culture can be related to class, where class, such as working class, is the dominant feature over ethnicity or other factors, as

Figure 8.1 National pride is an important part of spaces in Washington, DC: Here a parent photographs a child at the National World War II Memorial, with the Lincoln Memorial in the background

Source: Paul Tranter

Karsten (1998) found in her Amsterdam study. When exploring culture, place can be paramount; children can share more culturally with their neighbours than, for example, with children of the same cultural or ethnic background living elsewhere.

Culture is the way in which society expresses itself and what it values. For children, culture can be experienced positively, negatively or both in different parts of their lives. Cultures can differentially value children's needs; physical traits such as gender, coloration and abilities such as sporting ability; intellectual capacity; and personal character. Societies are becoming more mobile and more multicultural. Few urban children live in a monocultural society; as Thompson (2003) states: 'The contemporary Western city is one of capital "D" Difference, a Difference which is, at one end of the spectrum, celebrated and joyously embraced; and at the other, denied, mistrusted and challenged.' In this chapter we explore this process, examining how culture and belonging enriches children's lives, but can also hinder their lives where belonging becomes associated with prejudice, neglect and unequal access to resources. In cities today children's lives can be multifaceted as they negotiate a complex raft of cultural positions, attributes and beliefs. The fact that they manage this with assurance and tenacity is the theme of this chapter; it is one of hope as children move through and celebrate increasingly diverse cultural heritages in cities today. In this chapter we explore the role of culture in children's spatial lives, notably how culture relates to exclusion and inclusion, the ways in which culture is celebrated, and the spaces that permeate and support culture (see Figure 8.2).

Figure 8.2 Children interact with the city in diverse ways: Here, boys in Japan remember their ancestors

Source: Motohide Miyahara

Recognition of the Right to 'Culture' in Children's Lives

The position of children in cultural spaces has been little explored, yet is a critical element of their life world. The child's right to experience culture is enshrined in Article 30: 'Rights to name, culture and nationality' of the 1989 United Nations Convention on the Rights of the Child (UNCROC):

> In those States in which ethnic, religious or linguistic minorities or persons of indigenous origin exist, a child belonging to such a minority or who is indigenous

> shall not be denied the right, in community with other members of his or her group, to enjoy his or her own culture, to profess and practise his or her own religion, or to use his or her own language.

However, cultural rights need to be consistent with wider human rights:

> Every human being has the right to culture, including the right to enjoy and develop cultural life and identity. Cultural rights, however, are not unlimited. The right to culture is limited at the point at which it infringes on another human right. (Ayton-Shenker, 1995)

The right to practise one's culture is one that has also been recognized at national and local government level. In 1982, multiculturalism was recognized in the Canadian Charter of Rights and Freedoms. More recently, in 1997 the UK government made a commitment to a non-racist and multicultural Britain, and in Australia the government has expressed support for multiculturalism as part of its national education strategy.

Multiple Societies

Rising multiculturalism is one of the obvious changes taking place in cities today. Society is becoming more mobile as the internationalization and interdependence of national economies grows, transport and communications develop, and the creation of international organizations such as the European Union facilitates cross border migration. Some cities have historically been characterized by the presence of different racial and ethnic groups contributing to a rich cultural diversity, as indicated in the following examples: New York (African–American, Italian, Irish, German, Russian, Jewish, Polish), London (Jewish, Irish, East European, Afro-Caribbean, Pakistani, Indian and Bangladeshi) and Marseilles (Greek, Arab, Italian, Spanish, Berber, Algerian, Comoros, Chinese and Vietnamese). The groups identified for these cities are indicative and not intended to cover the full range of population groups present in these cities. During the 21st century, the diversity exhibited in New York, London and Marseilles is widespread and typically characteristic of most Western cities, rather than, as was the case historically, specific to a few cities that formed part of international trade routes. As Sandercock (2003, p319) writes:

The 21st century is indisputably the century of multicultural cities

> The 21st century is indisputably the century of multicultural cities. It will also be the century of struggle for multiculturalism and against fundamentalism, for tolerance and pluralisation.

The extent of this emergent diversity can be seen in the city of Toronto, which heralds itself as one of the most multicultural cities in the world, with over 140 languages spoken (City of Toronto, 1998–2010). As societies become more multicultural, each group will bring with it its cultural norms, religious and spiritual beliefs, its language and its ways of valuing childhood. But culture is not just about race, language or ethnicity. It is about ways of doing, values and beliefs. The rise of multiculturalism in cities means that children today will be exposed to, and will themselves contribute to, a much wider range of cultural ways of thinking and being than any previous society. Few city children today grow up in a homogeneous society. It is essential that they are prepared for life in a heterogeneous, diverse cultural mix, and as this chapter will show, it is children who are in many ways better equipped and open to embracing this new urban society than their parents' generation.

'D' for Difference: Cultural Debates

What cultural practices matter to children and what level of adherence to culture should be given? What happens when cultural norms are at odds with wider understandings of children's rights? An indicative example is schooling, where cultural expectations prioritize school achievement above play, leisure and family activities. Thus, children may have minimal time to play and relax; yet the right to both education and leisure and play are enshrined in the UNCROC. Other examples include cultural expectations around gender where these can inhibit (usually girls') rights to fully engage in education and may restrict their ability to independently access places in the city where boys can more easily go. Debates around culture and the place of culture become especially pronounced where expressions of culture are most visible, through dress, language, place of abode or physical features.

What happens when cultural norms are at odds with children's rights?

Children move between different cultural spaces in different parts of their lives. They internalize and manifest behaviours that are appropriate to different places, the school, the home, the place of worship, public meeting spaces and the street. They develop expertise at becoming urban chameleons to 'blend' in when facing the 'dominant' culture. For some children being chameleons is not something they wish to do and is harder for them than for adults: through their dress, language and physical features they carry with them their 'difference'. A child with a disability, or a girl child wearing a *hijab* or a boy child wearing a *yarmulke* will move between places where they can be part of the dominant culture, usually at home, school or place of worship (where girls and women commonly wear the *hijab*), to places where they are a minority culture and distinguished as such by their disability or aspects of their dress.

Population statistics ascribe people to a particular population, ethnic, religious or cultural group. For many children, such simplistic ascriptions are nonsensical. In his study of multicultural identities in a school classroom in inner London (27 nationalities), Joerchel (2006) explored in depth the multicultural worlds of three girls. Their parents represented different national groups (European, African and Middle Eastern), different religions (Christian and Muslim), different languages and had different cultural practices. The children, Joerchel (2006) found, exhibited a pride in their own 'difference' and were moving towards a balance between the various expressions of their own and the wider dominant culture. Through TV (accessing stations from their home country), parental support for, and pride in, their cultural background, and a positive supportive school environment, the children were able to embrace their difference. As cities diversify, more children will be born with multiple cultural, racial and language backgrounds. The critical factor, for children, is not whether they are dominant or a minority, not their ability or

Culture should not be a barrier to accessing societal resources

disability, nor their ethnicity or language, but the extent of acceptance and embracing of their culture and the provision of appropriate support for children as they move between cultures. Culture should not be a barrier to accessing societal resources or open children to prejudice or discrimination. The following section examines culture as a factor in exclusion.

Segregation and Inclusion: Race and Ethnicity

Historically, some cities have been infamous for the presence of racial and ethnic ghettos. Among the best-known examples are Harlem in New York (Afro-American), the Jewish ghettos in Vienna and Prague, as well as city neighbourhoods characterized by distinct population groups, such as the Chinese quarters in many major cities. Less noted are the often rich cultural lives of their residents. While the term 'ghetto' is largely obsolete in its original sense as an area where a particular cultural or population group is compelled to live, city neighbourhoods in which specific cultural, racial or population groups gather are still evident. Such areas are often associated with high levels of poverty and housing need, and score highly on deprivation indices. Living in poor and segregated neighbourhoods is potentially detrimental to children's well-being (Niles and Peck, 2008).

In cities, ethnic groups are not characteristically evenly distributed across space and can show substantial variations at both the regional and local level. Birmingham in the centre of England is indicative in this regard. The 2001 census statistics for the UK indicated that England, as a whole, had 9 per cent of its population classed as belonging to an ethnic minority, while in Birmingham 34 per cent of residents considered themselves to be 'non-white'. Of this 34 per cent, over half were found in 7 of the city's 39 wards (Birmingham City Council, 2001). A ward is an administrative area like a city suburb. Different wards have different concentrations of ethnic groups. In a study of the black public sphere and popular cultures in the city, Dudrah (2002) interviewed a number of young people who spoke positively of the inner city as a place of belonging, created over years of settlement and struggle, compared to the outer suburbs. Within the inner city, culture is openly expressed and supported through Asian shops with the latest fashions and jewellery, fresh spices and vegetables from around the world, Caribbean bakers, Asian sweet shops, street music emanating from Asian video and music stores, and collective celebrations such as *Vaisakhi* (when India flags are flown), *Eid* (with its Pakistani flags) and the Handsworth Carnival (Caribbean). Bharat, a local Birmingham coach company, provides coaches that link their mainly South Asian customers with South Asian communities in other parts of England. Cultural connections can 'pass over space' so that a child may have more connectivity (catching a Bharat coach) with a similar ethnic community in a different city than with a different ethnic community in their own city or neighbourhood.

The multicultural nature of Birmingham parallels that of Amsterdam, although Amsterdam's constituent ethnic groups are obviously reflective of its different historic heritage. The impact of segregation upon children's lives is the focus of Lia Karsten's (1998) study of Amsterdam, which, in common with many European cities, is strongly ethnically diverse, with the three largest ethnic populations being Turkish–Moroccan, Surinamese–Antillean and white Dutch. In the study, Karsten (1998) uses the term Dutch to refer to ethnic white Dutch, though ethnic children may also have Dutch nationality. By 1994 Dutch children comprised only 34 per cent of all Amsterdam children. The study identified several ethnic differences among children. With regard to family size, Turkish and Moroccan children tend to live in bigger families, thus having more access to playmates. In play, especially group-based playground play, the largest group in terms of gender and ethnicity appears to dominate and determine the rules. Thus, the study found that although Turkish and Moroccan children constituted 50 per cent of the overall neighbourhood population, they comprised 70 per cent of playing children at the playground. Similarly, in another neighbourhood where Dutch Amsterdam children comprised 50 per cent of the overall population, they accounted for 60 per cent of the playground population. This suggests that children belonging to numerically smaller populations in neighbourhoods can be disadvantaged in their access to play space. Also disadvantaged are girls, as two-thirds of the children observed at playgrounds were boys, with under-representation of girls occurring for all ethnic groups. Girls also spend less time at playgrounds. Furthermore, at the micro-scale of the play area, girls tend to inhabit the outer areas, while boys take the open spaces, mainly for soccer. Gender is indicative of the wider complexity of assessing children's experiences by reference to one characteristic, in this case ethnicity. For as Karsten (1998) concludes, while at first sight ethnic divides are apparent, gender and class analysis reveals a much more complex picture. Divides along cultural, ethnic and religious lines are particularly evident in schools, and raise challenging issues for parents and society on matters of assimilation/integration and separation.

In play, the largest group in terms of gender and ethnicity dominates and determines the rules

Segregation and Inclusion: Schools

Segregation at the school level can be more distinct than in the population generally, as Rangvid (2007) found in Copenhagen and Karsten (1998) found in Amsterdam. Schools in the UK, the US, and parts of Europe and Australasia can exhibit what is pejoratively termed 'white flight', where schools are characterized by the absence of white ethnic groups. In parts of South Auckland, New Zealand's largest city and often also referred to as the number one Pacific city due to its large Pasifika population, schools can exhibit very high levels of Pasifika children. Pasifika children can constitute 90 per cent of the total school population, proportions not necessarily representative of the wider community mix. Moreover, in 2003, 68 per cent of the Pasifika school population were in decile 1, 2 or 3 schools, where deciles work on a scale of 1 to 10, with 1 being the lowest socio-economic status. This compares with only 26 per cent of the total school population being in decile 1, 2 or 3 schools (Ministry of Education, 2008). The ethnic diversity and high Pasifika school rolls can be seen in the statistics for three primary schools in South Auckland (see Table 8.1).

What this means is that children in these high ethnic minority schools can be experiencing less ethnically diverse populations than their adult relatives, making it harder for children to experience the full ethnic diversity of their wider community or to mix with children from the dominant nationality. On

Table 8.1 Population composition for three schools in South Auckland, New Zealand

	School 1	School 2	School 3
School population composition	• Boys: 50%; girls: 50% • Ethnic composition: New Zealand Māori: 26%; New Zealand European/ Pākehā:* 21%; Samoan: 20%; Tongan: 8%; Cook Island Māori: 6%; Indian: 6%; Chinese: 3%; Niuean: 3%; Fijian: 1%; other Asian: 1%; other European: 1%; other: 4% (Māori Pasifika = 64%)	• Girls: 53%; boys: 47% • Ethnic composition: Māori: 18%; New Zealand European/Pākehā: 1%; Samoan: 37%; Tongan: 22%; Cook Island Māori: 16%; Niuean: 5%; Tokelauan: 1% (Māori Pasifika = 99%)	• Girls: 53%; boys: 47% • Ethnic composition: Māori: 18%; New Zealand European/Pākehā: 1%; Samoan: 47%; Tongan: 19%; Cook Island Māori: 12%; Niuean: 2%; other: 1% (Māori Pasifika = 98%)
Local area population composition	• Pasifika: 31%; Māori: 21%; European: 48%; Asian: 12%; other: 7%**	• Pasifika: 68%; Māori: 20%; European and Chinese: 11%	• Pasifika: 68%; Māori: 20%; European and Chinese: 11%

Notes: * Pākehā = white European; Māori are the indigenous people of New Zealand.

Unfortunately, the two sets of figures are not directly comparable: in the census people can belong to more than one ethnic group, so totals will not amount to 100%.

Source: adapted from Ministry for Education (2008); Statistics New Zealand (2006)

the positive side, such schools often have strong programmes supporting children's cultural identities (see Figure 8.3). In school, regardless of the ethnic affiliations of the children, the curriculum is one that will invariably be representative of the 'dominant' national group, especially in the history curriculum (Amin, 2002).

Figure 8.3 Schools can play an important role in supporting children's cultural identities through activities such as dance: Pasifika children performing at a school in Auckland, New Zealand

Source: School of Architecture and Planning, Auckland University

Perhaps the best-known attempt to redress issues of racial segregation in schools was the US Bus Initiative. Starting in the early 1970s, children were bussed to schools of a different ethnic group to achieve a 'balance' and to move towards greater equality of educational opportunity. The extent of the segregation problem in US inner-city public schools is indicative of the fact that as recently as 2003, the majority of the population in public schools (75 per cent for the 25 largest cities in the US) is made up of minorities (Wilson, cited in Welch, 2007). In racial terms, some of the results from the bussing programme were impressive: 'the percentage of black students attending schools that were 90 to 100 per cent minority fell from 58 per cent in Charlotte and 66 per cent in Louisville to 2 per cent in both districts immediately following the rulings' (Clotfelter, cited in Welch, 2007, p55). The numbers of children bussed were substantial, as indicated by the fact that for 19 years, St Louis bussed over 13,000 black city students annually to white-majority schools in over 20 suburban school districts. A national survey of public school teachers' and students' opinions of school-related interracial contact reported that over 90 per cent of the teachers and over 80 per cent of the students polled said that attending classes, socializing and participating in after-school activities with students of differing racial and ethnic backgrounds was important (Welch, 2007, p56).

Perhaps the best-known attempt to redress racial segregation in schools was the US Bus Initiative

Where extreme segregation is occurring, the Bus Initiative, which is in itself an extreme response, may have some validity. However, while the children may experience benefits, there are also dis-benefits. Children leave their home neighbourhood, have reduced contact with their own cultural background and may experience the disbenefits of being bussed into a situation where they become the minority population. In the US, school bussing is now ending and schools are again resegregating. The problem in American schools is not an unusual one; but more than race and culture, the problem is one of uneven distribution of resources, with schools in predominantly 'ethnic' neighbourhoods perceived as being of lesser quality and associated with poorer achievement levels. Cultural or racial difference can be a 'red herring', diverting attention from persistent underlying social and economic inequalities. The issue of segregation is not limited to ethnicity or race, but is highly pertinent to children of differing abilities.

Segregation and Inclusion: Disability

One seldom discussed area of separation is that of differentially abled children – that is, children with a physical and/or mental disability – which is especially common in the school context. Children with a disability frequently experience exclusion in a number of areas of their lives in addition to the challenges already experienced as a result of their disability. Most countries have legislation and policies designed to recognize and support children with disabilities, in line with the United Nations Educational, Scientific and Cultural Organization (UNESCO) 1960 Convention against Discrimination in Education and, more recently, the 1994 Salamanca Statement and Framework for Action on Special Educational Needs, which supports inclusive education. The majority of disabled children can, with support, take part in mainstream education. Yet, only 59 per cent of children in a seven-country European-funded study were found to do so, with 12.4 per cent receiving no education. Only 10 per cent of respondents reported that the support provided was sufficient (European Disability Forum, 2003). Of particular concern is that 73 per cent of respondents noted that no support is provided for extra-curricular activities. This means that children are less able to participate in activities that take place outside the main classroom, missing out on what are usually the more fun and challenging aspects of their schooling. Different countries respond differently to where children receive their main school education – that is, in mainstream schools or separate schools. The UK is one country that has recently moved towards greater integration of children with disabilities into mainstream schools (Holt, 2003). In New Zealand, there is a long-established culture of educating disabled children in mainstream schools. However, if such children are educated in mainstream schools, their educational needs must be supported, and the environment should be responsive to their physical and social needs, such that mobility-impaired children, for example, are able to move around easily. Placing children in a space that is not appropriate to meeting their intellectual or physical needs can enhance their feelings of exclusion and alienation. Playgrounds can be especially isolating places for disabled children (Yantzi et al, 2010; see also Chapter 7).

Playgrounds can be especially isolating places for disabled children

The issues confronting differentially abled children can be illustrated by reference to those experienced by deaf children. Deaf children experience marginalization in a number of areas of their lives. Separation into special schools enhances marginalization; but even where children are physically in the same space in mainstream schools, they can be outsiders as they struggle to communicate with the majority population. Some 95 per cent of D/deaf people (a term used to cover all people with some type of deafness and includes those who are hard of hearing, partially deaf or profoundly deaf) who see themselves as part of the deaf community are born into hearing families (Valentine and Skelton, 2003). Their disability is seen as a medical problem, with the aim being to provide them with the means wherever possible of integrating them into 'normal' (i.e. hearing) society through hearing technologies and medical interventions, such as cochlear implants. Children born into non-hearing families and who become fluent in signing have enhanced access to Deaf identity and culture. In this sense, children who are taught to sign in a language different from their home culture can be particularly marginalized, such as deaf Asian children who are taught British sign language (BSL), which can exacerbate their feelings of disconnection from their family and culture as they can still find it difficult to participate in events and celebrations at family and community level.

Segregation and Inclusion: Religion

The segregation of children in schools is not a new concept and one not only common in private schools, which separate along economic lines, but also evident in religious-affiliated schools. In the UK, of 1,197,000 primary school pupils, nearly 36 per cent of primary aged children attending a government-funded school attend a faith-based school (Bolton and Gillie, 2009). In the US, private schools are attended by 25 per cent of children, with around 80 per cent of these having a religious affiliation (National Centre for Educational Statistics, 2008).

Where children attend such schools, there is an increased likelihood that they will travel longer distances to schools and have more of their social contacts outside their home neighbourhood. Debate has been especially strong on the issue of religion and culture as it concerns Islamic-based faith schools, and is unusual in the sense that there has been long acceptance of faith-based education for other schools affiliated to Catholic, Church of England and Jewish religions. In France, a law passed in 2004 specifically prohibits the wearing of symbols or clothing that show religious affiliation in public primary and secondary schools. This law impacts directly upon those Muslim schoolgirls who wear headscarves. Muslim girls in their daily lives move between places where physical expressions of their religion/culture are spatially variable in their acceptability. The issues around Islamic schools have been heavily debated in the popular press, and can be especially stressful for Muslim parents and children. Muslim communities can find that they reside within culturally incongruent places, and thus seek to shelter their children from negative outside influences by sending them to culturally supportive schools (Zine, 2007, p71). Islamic schools are gaining support from Muslim parents as they are seen to support Islamic cultural and religious beliefs, provide safer environments and provide better educational environments for their children, who often fail to achieve in mainstream schools. Attending a Muslim school can be beneficial for children in practical terms as it overcomes the need for them to attend additional after-school classes, deemed necessary to maintain children's cultural and religious knowledge. These classes can be long and tiring for children on top of a full school day. In Canadian cities, there has been a rapid growth in the number of Islamic schools. Some ten years ago, Toronto had 18 (the province of Ontario had 35) such schools, with the number continuing to grow since (Zine, 2007). The same trend is evident in the UK, which in 2008 had around 110 private Islamic schools, 9 of which were government-funded/maintained schools.

The youthful demographic for Muslim populations means that there are higher numbers of Muslim children in schools in proportion to their overall population. Muslims also are ethnically diverse. In the UK, the main Muslim groups are Pakistani (40 per cent); Bangladeshi (20 per cent); Turkish, Turkish–Cypriot, Middle Eastern, East Asian and African-Caribbean (10 per cent); Indian or other South Asians (15 per cent); mixed race (4 per cent); and white converts and Eastern Europeans (1 per cent) (Burgess et al, 2005). This diversity is significant as one of the criticisms directed at Islamic-based schools is that they are exclusionary. This diversity is also present in the backgrounds of the children in Figure 8.4, a photo taken in New Zealand mosque. Whatever the arguments and philosophical positions for and against segregated Islamic schools, there is no doubt that they are growing and are an important part of the educational and life context of many Muslim children in European cities. As part of our intention to include children's voices in this book, children (aged 6 to 14) who attend a mosque in New Zealand were asked about their experiences of the mosque. Their responses are summarized in Box 8.1 and are interesting in that although the children recognize the mosque as being primarily a place of prayer and learning (the *Quran*), it is also described as a fun place, a community, a place of safety and belonging and where the children go to meet their friends. The mosque is a highly valued and important space in their lives.

Children describe the mosque as a fun place, a community, a place of safety and belonging

Figure 8.4 These children attending a mosque, although all Muslim, represent a range of ethnic, language and cultural groups

Source: Mai Tamimi

Box 8.1 Children's experiences in a mosque

The mosque to me is where I worship and where I am a lot of the time. I like the mosque mainly because of the community: everyone knows and likes each other, like a big family.

I like the mosque because we pray at the mosque and we can learn something about the *Quran*. I love to go to the mosque and I like learning the *Quran*.

I learn about Islam. I learn more about prayer. I make new friends at the *Masjid*.

I read the *Quran*; I catch up with my friends.

I come here to learn about the *Quran* and learn more about Islam. So many special things happen in the *Masjid* that doesn't happen elsewhere: fun days for *Eids* and the start of Ramadan.

I love the house of Allah; it's fun.

Fun, people with the same religion, makes you feel at home.

I like this place because I have nice friends and I learn lots.

I like this place because I like to come and pray. I like to break my fast.

I like Sunday school, especially the teachers.

I like *Masjid* because it's fun; we get prizes and nice friends

This mosque is quite good, maybe the best mosque I've been in my entire life. It's a really great mosque, having the chance to learn about Islam and life, peace and patience.

The mosque to me is the place where I worship Allah; it makes me feel happier than other places. The mosque is a place to learn more about my religion. It's a peaceful place and I feel safe in it.

Segregation and Inclusion: Indigenous Children

Similar but less acrimonious concerns about children and culture have been raised with regard to indigenous children. These children typically score less well on standard educational attainment tests, leave school earlier and are in environments that are seen as linguistically and culturally failing to support them.

Indigenous children in cities share many of the problems experienced by children of immigrant ethnic minority families: poor health, poor educational attainment, exposure to racism and prejudice, and difficulties in maintaining their own cultural identities, practices and language. In addition, indigenous communities also face additional problems emanating from the process of colonization and its associated struggles. Indigenous children, particularly in the US (Native Americans), Canada (Inuit, First Nations), Australia (Aboriginal) and New Zealand (Māori-Pasifika), also score low on well-being indicators. A major study looking at the health of indigenous children in the last three of these countries found profound and unjust differences in child health between indigenous children and the wider child population, and low educational achievement (Smylie and Adomako, 2009).

Indigenous children share many of the problems experienced by children of immigrants

The example of Māori, the indigenous peoples of New Zealand, illustrates how a focused approach can make huge impact. During the 1980s there was what Hingangaroa Smith (2003) calls a 'revolutionary mind shift' among the Māori away from reactive and towards proactive action. The needs of Māori children in urban areas were seen as pressing, as they were failing educationally and becoming removed from their culture, which is traditionally rural based. Māori language was also under threat. During the 1980s, this concern inspired the *Te Kōhanga Reo* (early childhood centres) initiative. *Te Kōhanga Reo* is a Māori development initiative, inspired by Māori elders in 1982, and aimed at maintaining and strengthening Māori language and philosophies within a cultural framework (see the *Te Kōhanga Reo* National Trust website at www.kohanga.ac.nz). Children from birth to the age of six are immersed in Māori language and culture. The *Kōhanga Reo* were quickly followed by *Kura Kaupapa* (Māori immersion schools at both primary and secondary level). These schools support Māori culture, language and aspirations on a *whanau* (family) basis, where children invest not in themselves alone, but in the *whanau* of which all are members. This approach has resulted in parents who found their own educational experience alienating now reinvesting in schooling and education for their children (Hingangaroa Smith, 2003). The success of the Māori educational initiative has been recognized in a United Nations report on *Reaching the Marginalized* (United Nations, 2010, p30). Like Islamic schools, there has been a substantial growth in *Te Kōhanga Reo* and *Kura Kaupapa* schools – now at 73 *Kura Kaupapa* and over 800 *Kōhanga Reo* in New Zealand. Not all are in cities; but it is in cities where they provide children with access to their culture and language that may otherwise be difficult to find in an urban environment often removed from their usually non-urban based *marae* (meeting house) and ancestral homes.

Ethnicity can be a particularly fraught issue, with polarized views between perceiving higher numbers of ethnic groups in identifiable spatial locations as a positive support for culture or as inhibiting cultural sharing and integration. Spatial concentration can be seen as providing safety from racist and ethnic attacks and providing access to culturally supportive institutions such as temples, shops and language schools. Conversely, ethnic groupings are seen as inhibiting cultural sharing and children's well-being (notably, girl children in cultures perceived to have restrictive gender roles), and, in the case of migrants, as a threat to the dominant culture and precluding children's access to wider social networks and resources. Next we look at the case of Northern Ireland: an extreme example of religious affiliations affecting children's lives.

Segregation and Conflict

Much has been written on children and risk, identifying places in which children feel safe and unsafe. In many cases, parental fears for their children's safety may be exaggerated compared to the actual risks. However, in the following example, the perceived risks may be closer to actual risks. The example used here looks at Northern Ireland, where children's perceptions and experiences of safety are closely connected to their religious (i.e. Roman Catholic or Protestant) identity. Northern Ireland presents a severe and disturbing illustration of the impact of historic and ongoing conflict and its associated segregation upon children's lives:

> There is growing evidence that the conflict has had a traumatising effect on far larger numbers of children and young people than was formerly acknowledged ... Growing up in a highly segregated society like Northern Ireland has been shown to also take its toll. Segregation affects every aspect of children's lives: housing, education and leisure. (Northern Ireland Human Rights Association, 2006)

Children living closest to the interface between the two religious communities felt most unsafe

A peace settlement was reached on 10 April 1998, known as the Good Friday Agreement – resulting in a ceasefire; however, the effects of conflict and segregation continue to affect children years after. A study undertaken five years after the peace agreement found that young people's feelings of safety were strongly influenced by where they lived. The validity of children's fears is evident in light of the fact that as of March 1998, 3601 people – of whom 257 were aged under 18 – had died in the conflict (Smyth, 1998). Deaths and injuries were highest in areas of most extreme deprivation. Many of the children's families had been, and continue to be, directly affected by the legacies of the 'Troubles', as the conflict is known. Children aged 14 and living closest to the interface between the two religious communities felt most unsafe. Pupils from three of the four schools involved in the study defined their homes as unsafe, especially when on their own or at night (Leonard, 2007).

Fear was also important in determining how space was used to access place; thus, shops on the periphery were avoided, as were shared social spaces such as leisure facilities located in areas seen as belonging to a different religious group. In Northern Ireland, spaces often carry clear indications of identity through wall murals and pavement painting – for example, Protestant areas can have the edges of the pavements painted red, white and blue. Interestingly, in contrast to findings elsewhere, Leonard (2007) found that girls were more spatially mobile, a finding related to the fact that boys are more likely to encounter and be victims of intercommunity violence. Children adopted a number of strategies to deal with risk. These included:

- risk avoidance – avoiding areas belonging to the other religion;
- risk minimization – not using backyards if living close to the 'peace' lines (barriers separating the two religious communities);
- risk management – concealing religious identity when using shared spaces;
- risk taking – actively occupying contested spaces.

Figure 8.5 Children in traditional dress taking part in the St Patrick's Day Parade in Derry, Northern Ireland

Source: Claire Freeman

Children in Northern Ireland live in one of the UK's most deprived areas. Although the risks faced by Northern Ireland children are higher than those that children are normally exposed to, the process of avoiding, confronting, negotiating and renegotiating space are processes common to children across cities and societies where safety can be quite variable across both time and space. The Northern Ireland example is one where religion acts as a divide and has negative impacts upon the life of some of its most vulnerable children. On the positive side, religion and culture can also be a cause of celebration and reassurance to children of their culture. The photo of the St Patrick's Day Parade in Derry, Northern Ireland, shows how even societies with their difficult history can rise above conflict and celebrate (see Figure 8.5).

Celebrations, Observances and Festivals

In discussions of culture and difference, it is too easy to overlook the fact that for the majority of children, religion, race, ethnicity and culture are valued for their distinctiveness, diversity and the way in which they shape better ways of seeing, experiencing and interacting with the world. In trawling the literature for this chapter, it was interesting to note that while there were innumerable studies on children and race, segregation and 'problems' associated with culture, there was minimal information available on children and cultural celebration, and culture as a unifying force for children in society. This comparative silence is especially surprising in light of the growing multicultural nature of urban life.

In multicultural societies, children are becoming exposed to, and increasingly taking part in, a range of festivities and cultural observances and commemorations. Children in larger cities commonly experience, through city celebrations or interactions with school friends, *Divali* (Hindu), *Eid* (Muslim), the Chinese New Year, Christmas and Easter. Although they may not necessarily celebrate themselves, they will be aware of these occasions through seeing Christmas lights, through hearing fireworks and seeing Chinese dragons during the Chinese New Year, watching carnivals and parades, and through increasingly multicultural social education in schools. An indicative selection of cultural events is listed in Table 8.2. These events exhibit some surprisingly strong commonalities: feasting and making special foods; the coming together of extended families; recognition of life events and life phases; birthdays, marriage and death; and the changing seasons. Events which have a serious purpose, such as the Day of the Dead (Mexican), Yom Kippur (Jewish) and Ramadan (Muslim), are also associated with feasting and coming together of families, and the affirmation of life. Some events specifically focus on children, such as *Tol* (Korea), birthdays (see Figure 8.6) and *Bar/Bat Mitzvah*; but while most do not single children out, they do provide additional recognition of children's role and place in normally adult-dominated life worlds. Examples are given in the last column of Table 8.2, where children take a more obvious role, as in *Eid*, where children are encouraged to dress in traditional clothing and take a more prominent part in the prayers than is usual, and later take part in the feasts. Thus, during times of celebration or observance, what can be seen is a higher level of welcome, tolerance and recognition for children than is usual in a community or group's sacred and culturally important spaces.

Figure 8.6 Birthdays are an important celebration in many children's lives, ranging from modest parties at home with friends and family, as in this photo, to larger expensive affairs held in public venues

Source: Claire Freeman

Table 8.2 Culturally significant events

Festival	Name and place			Children's role
Religious	*Eid-ul-Fitr** and *Eid-ul-Adha*: Islam	Christmas and Easter: Christian	*Maslenitsa*: Russia	*Eid-ul-Fitr*: children go with their parents to the mosque to worship and celebrate the end of Ramadan; gifts can be exchanged; everyone joins in celebratory feasts.
New Year	Matariki/Māori New Year	Chinese New Year*	Hogmanay: Scotland	Chinese New Year: families gather together to celebrate and feast. Children receive a red envelope with money and sweet delicacies. Includes lighting lanterns, fireworks and dragon dances.
Changing of the seasons	*Walpurgis* and *Midsommar*: Scandinavia	Cherry Blossom Festival: Japan*	*Sankt Hans Aften*: Denmark	Cherry Blossom Festival: families gather and hold picnics under the cherry trees.
Cultural/ immigrant/ community festivals	Pasifika festivals: New Zealand	Notting Hill Carnival: Caribbean community, UK*	*Mela*: Europe, especially UK, and South Asia	Notting Hill Carnival: first day of the carnival is children's day; children, like the adults, don costumes and parade, supported by special children's and family events.
National celebrations	Australia Day	Independence Day: US*	*Sami* National Day: Norway	Independence Day: celebrated with fireworks, picnics, baseball games, concerts and public events attended by families.
Remembrance	Thanksgiving: Canada and US	*Pesach*/ Passover: Jewish*	Obon Festival: Japan	*Pesach*/Passover : the *seder* meal begins with the youngest child asking the traditional question; this is followed by a meal and a period of prayer and fasting.
Lights/colours	*Diwali*:* Hindu	*Hannukah*: Jewish	Lantern Festival: China	*Diwali*: a five-day festival celebrating the triumph of good over evil; families celebrate with fireworks, lights and feasts
Leaders: figures of significance	Emperor's birthday: Japan	Dragon Boat Festival: China	*Vesak/Viskaha Puja*/Buddha's birthday:* east Asia	*Vesak/Viskaha Puja*: celebrates the teachings of Buddha with visits to the temple and offerings; processions can include dragons and lanterns. Children are central to the day as they carry on his legacy.
Historic occasions	Guy Fawkes/ Bonfire Night*: UK	Bastille Day: France	Orange Day: The Netherlands	Guy Fawkes: celebrated with fireworks and bonfires. Children traditionally make a 'guy' (dummy) to go on the fire and collect money with their 'guys'.
Age/life-phase celebrations	Birthdays: various countries	*Bar Mitzvah* and *Bat Mitzvah*: Jewish	*Shichi-go-san*:* Japan	*Shichi-go-san*: rite of passage and festival day for three- and seven- year-old girls, and three- and five-year-old boys; children wear traditional dress, visit the temple and also get 1000-year candy.
Place/local festivals	Mardi Gras:* New Orleans, US	*La Fête Nationale du Québec*: Canada	Carnival of Venice: Italy	Mardi Gras: an adult festival but with designated family-friendly parade venues; families line the streets to watch the parades; some children sit on special raised seats attached to ladders.
Other	International children's days: speeches, public events and family activities	Halloween: children dress up at night and go trick or treating, collecting sweets from houses	Mischief night – children play tricks and get up to mischief in the neighbourhood: northern England and parts of the US	Birthdays: in developed countries, mostly as individual celebrations; usually associated with presents, parties, family gatherings and age group gatherings. Celebration may take place at home or in the wider community.

Note: * This event is further detailed in the final column in terms of children's role.

Internationalization of Culture

The earlier part of this chapter focused mainly on culture as a point of difference; however, one of the most important trends culturally in children's lives has been the promotion of universal cultural elements that transcend space and other 'cultural' differences, such as language, nation, ethnicity and geographic location. The most obvious examples are in the spread of items relating to consumerist culture, particularly to toys. The term 'Disneyfication' has been used to refer to the spread of toys associated with Disney films and which are encouraged through their associations with outlets such as McDonald's. The development of technology now means that televisions, DVD players, iPhones, PlayStations and Xbox are available to, and used by, children internationally. This means that children can transcend their own spatial and cultural limitations when engaging with such technology and in playing games remotely. This process will be self-evident to anyone who has observed children travelling to different countries, where one child playing a hand-held game will often attract other children who, despite different languages, ages and experiences, are able to connect through the game. Games can, therefore, act as an important entry point for children to access other children who may be very different from themselves. On a less technological level, a soccer ball can achieve the same result.

Children transcend spatial and cultural limitations when engaging with technology

The universalization of culture can also be seen in the globalization of building form, such that a child in Hong Kong, Seattle (US), Madrid (Spain) and Canberra (Australia) may live in architecturally similar apartments, use very similar shopping malls and be driven along similar looking roads in near-identical vehicles. Another unifying factor is that of sport. While some sports are specific to certain countries, such as Aussie rules football, American football, skiing and sumo wrestling, others (most obviously soccer) are common to children across the globe. Not only do children play soccer all over the world, but support for teams can transcend space, as with Japanese children wearing Manchester United soccer shirts or English children wearing Real Madrid or Brazil shirts. Clothing in its wider sense also transcends space, such that children internationally will commonly wear T-shirts and jeans, whether living through a cold Canadian winter or a hot Spanish summer. Ethnic food has become universally available. Rice, chips/French fries, sushi, pizza, bananas, curry and noodles, for example, which were once only available in their countries of origin, are now familiar to most urban children. Finally, some celebrations and festivals have become increasingly international. Birthdays and birthday parties, weddings, Christmas and New Year celebrations are all becoming commonplace in different countries, and part of a growing universal culture. Different cultures wear similar white wedding dresses, birthday parties occur in McDonald's worldwide, and even countries such as Australia still have fir trees and reindeer on their Christmas greeting cards. Children growing up in the 21st century will be familiar with international cultural items. Many will travel between countries and this familiarity can ease their entry into their encounters with different societies, whether through holidays, exchanges, migration or transnational adoptions. The world and its cultural diversity is becoming an experience open to children in cities, and one that they can experience by travel; but, more importantly, one that they can experience within their own city in their daily lives.

A Culturally Supportive City

This book is about the spaces that children inhabit and how these spaces shape their lives. Culture is increasingly transcending space such that children and their families can experience and benefit from the rich and diverse cultural worlds in which they live while living in their own cities. Between 2001 and 2006, Toronto welcomed 55,000 new migrants annually, with 47 per cent of residents having a mother tongue that is not English or French (City of Toronto, 1998–2010). Yet, Toronto, rather than being the exception, is indicative of trends in cities globally. Cities are multilingual, multi-ethnic, multiracial, multi-religious and multicultural. Given that globalization trends will encourage and possibly enhance international migration and exchange, and the fact that in many cities migrant populations are younger; cities will become 'more' not less diverse. The issue is how will cities respond to this 'difference'? As Sandercock (2000) attests, a built environment responsive to difference can be found in the interdependent concepts of 'right to the city' and the 'right to difference'. To date, cities have not adequately recognized children's rights, and neither have they been equitable in valuing difference, as is evident in the exclusion experienced by children seen as 'different'. Exclusion can be associated with space, as in segregated neighbourhoods, but can also take place in the same space where some children are able to colonize and utilize spaces better than others. Children with disabilities, or children unable to speak the majority language, are particularly vulnerable. On the positive side, children are leading the way: they are growing up with awareness of other cultures, with access to ways of thinking, being and playing that provide a common ground for cultural exchange that is often far more developed than that of adults. When new migrants come to a country, it is usually their children who adapt quickest, who help their parents traverse the myriad of language and cultural expectations that they confront. A city is only truly cultural when it acknowledges, values and shares all of its cultures, and when all spaces are accessible regardless of 'difference'. As yet, cities have a long path to tread to reach this state.

When new migrants come to a country, their children adapt quickest

References

Amin, A. (2002) *Ethnicity and the Multicultural City: Living with Diversity*, Report for the Department of Transport, Local Government and the Regions and the ESRC Cities Initiative, University of Durham, Durham, UK

Ayton-Shenker, D. (1995) *United Nations Background Note: The Challenge of Human Rights and Cultural Diversity*, United Nations, www.un.org/rights/dpi1627e.html, accessed March 2010

Birmingham City Council (2001) *2001 Population Census in Birmingham, Cultural Background: Ethnic and Religious Groups*, Planning Information Group, Development Directorate, Birmingham City Council, Birmingham, UK

Bolton, P. and Gillie, C. (2009) *Faith Schools: Admissions and performance, Standard Note, SN./SG/4405*, House of Commons Library, www.parliament.uk/briefingpapers/commons/lib/research/briefings/snsg-04405.pdf, accessed 11 January 2011

Bray, A. and Gates, S. (2000) 'Children with disabilities: Equal rights or different rights?', in A. B. Smith, M. Gollop, K. Marshall and K. Nairn (eds) *Advocating for Children*, University of Otago Press, Dunedin, New Zealand

Burgess, S., Wilson, D. and Lupton, R. (2005) 'Parallel lives? Ethnic segregation in schools and neighbourhoods', *Urban Studies*, vol 42, no 7, pp1027–1056
City of Toronto (1998–2010) *Living in Toronto: Toronto's Racial Diversity*, www.toronto.ca/toronto_facts/diversity.htm, accessed March 2010
Dudrah, R. K. (2002) 'Constructing city spaces through black popular cultures and the black public sphere', *City*, vol 6, pp335–350
European Disability Forum (2003) *Disability and Social Exclusion in the European Union: Time for Change, Tools for Change, Final Study Report*, European Disability Fund, European Commission, DG Employment and Social Affairs, http://cms.horus.be/files/99909/MediaArchive/pdf/disability%20and%20social%20exclusion%20in%20the%20eu.pdf/disability%20and%social%20exclusion%20in%20this%20eu.pdf
Hingangaroa Smith, G. (2003) *Kaupapa Maori Theory: Theorizing Indigenous Transformation of Education & Schooling*, University of Auckland & Te Whare Wananga o Awanuiarangi Tribal University New Zealand Kaupapa Maori Symposium, NZARE/AARE Joint Conference, Hyatt Hotel, Auckland, New Zealand, December 2003
Holt, L. (2003) '(Dis)abling children in primary school micro-spaces: Geographies of inclusion and exclusion', *Health and Place*, vol 9, pp119–128
Joerchel, A. C. (2006) 'A qualitative study of multicultural identities: Three cases of London's inner-city children', *Forum: Qualitative Social Research*, vol 7, no 2, Article 18
Karsten, L. (1998) 'Growing up in Amsterdam: Differentiation and segregation in children's daily lives', *Urban Studies*, vol 35, pp565–581
Leonard, M. (2007) 'Trapped in space? Children's accounts of risky environments', *Children and Society*, vol 21, pp432–445
Ministry of Education (2008) 'Pasifika Education Overview', www.minedu.govt.nz/NZEducation/Educationpolicies/PasifikaEducation.aspx, accessed March 2010
National Centre for Educational Statistics (2008) *Characteristics of Private Schools in the United States: Results from the 2007–08 Private School Universe Survey*, www.nces.ed.gov
Niles, M. D. and Peck, L. R. (2008) 'How poverty and segregation impact child development: Evidence from the Chicago longitudinal study', *Journal of Poverty*, vol 12, no 3, pp306–332
Northern Ireland Human Rights Association (2006) *Submission 281*, Save the Children and Children's Law Centre, http://billofrights.nihrc.org/search.asp?zoom_query=segregation
Rangvid, B. S. (2007) 'Living and learning separately? Ethnic segregation of school children in Copenhagen', *Urban Studies*, vol 44, no 7, pp1329–1354
Sandercock, L. (2000) 'Cities of (in)difference and the challenge for planning', *DISP*, vol 140, no 1, pp7–15
Sandercock, L. (2003) 'Planning in the ethno-culturally diverse city: A comment', *Planning Theory & Practice*, vol 4, no 3, pp319–323
Smylie, J. and Adomako, P. (eds) (2009) *Indigenous Children's Health Report: Health Assessment in Action*, Centre for Research on Inner City Health, Toronto, Canada
Smyth, M. (1998) *Half the Battle: Understanding the Impact of 'the Troubles' on Children and Young People*, The Cost of the Troubles Study Ltd 1996 to 1999, CAIN (Conflict Archive on the Internet), www.cain.ulst.ac.uk/
State of Queensland (2009) 'Module 6: Identity and culture', Department of Communities, The State of Queensland, Australia, www.childsafety.qld.gov.au/
Statistics New Zealand (2006) *Census Data: Ward Profiles Otara and Mangere Bridge*, New Zealand

Te Kōhanga Reo National Trust (2010) *Te kohanga reo*, www.kohanga.ac.nz/
Thompson, S. (2003) 'Planning and multiculturalism: A reflection on Australian local practice', *Planning Theory & Practice*, vol 4, no 3, pp275–293
United Nations (2010) *Reaching the Marginalized*, United Nations, UNESCO Publishing and Oxford University Press, Oxford
Valentine, G. and Skelton, T. (2003) 'Living on the edge: The marginalisation and "resistance" of D/deaf youth', *Environment and Planning* A, vol 35, pp301–321
Welch, P. J. (2007) 'How can US schools desegregate after the end of busing', *Forum for Social Economics*, vol 36, pp53–62
Yantzi, N. M., Young, N. L. and Mckeever, P. (2010) 'The suitability of school playgrounds for physically disabled children', *Children's Geographies*, vol 8, no 1, pp65–78
Zine, J. (2007) 'Safe havens or religious "ghettos"? Narratives of Islamic schooling in Canada', *Race Ethnicity and Education*, vol 10, no 1, pp71–92

9
Natural Space

> The enormous challenge facing us is how to minimize and mitigate the adverse environmental impacts of the modern built environment and how to provide more positive opportunities for contact with nature among children and adults as an integral part of everyday life. (Kellert, 1997, p89)

Nurturing a Love of Nature

London is not a city that immediately springs to mind when looking for a place in which to nurture world-renowned conservationists; yet this is where three such conservationists spent all or a significant part of their childhood. David Bellamy grew up in, and still lives in, London. He describes a field near his childhood home as a place in which all sorts of plants and animals could be seen, including adders (a venomous snake) and small deer, and a place that sparked his interest in nature (Bellamy, 2002). When asked in an interview about his childhood, David Attenborough, best known for his 'whispering' wildlife film commentary, recollects that as a boy living in London he collected natural specimens: 'I collected lots of things, fossils, snakes and salamanders and fish, and so on. I don't think I was exceptional' (Attenborough, 2009). A third conservationist growing up at least partly in London was Jane Goodall, the famous 'chimpanzee biologist' who wrote of her childhood:

Conservationists indicate the central importance of touching, feeling and observing wildlife as part of children's daily lives

> When I was a child, I spent as much time out of doors as I possibly could. I was fortunate: even when we lived in London there was a garden nearby so that I could spend time each day with grass and trees, small living creatures, and fresh air ... Most important, I had a remarkable and understanding mother. Once, she found me (aged 18 months old) watching in fascination the movements of a handful of earthworms I had taken up to my bed. Instead of expressing horror and demanding I throw them out, she quietly told me they would be dead if they could not get back to the earth. I ran with them into the garden. (Goodall, 1996)

These natural experiences, formative in the lives of these now internationally regarded conservationists, indicate the central importance of direct experience, touching, feeling and observing wildlife as part of their daily childhood lives. This chapter examines the lives of children in cities today, where and how they get their natural experiences, have opportunities to see and handle wildlife, and to spend time outdoors with nature. It addresses the following questions. If Bellamy, Attenborough or Goodall grew up in a big city such as London, Sydney, Tokyo or Rome today, could they still interact with nature in a way that engenders a lifelong love of things natural? Are natural experiences available to urban children? Can children growing up in cities today collect bugs, catch tadpoles, climb trees, feel the wind, rest in the shade of a favourite tree, throw stones, fight with sticks, pick flowers or play in the mud (see Figure 9.1)? If they can't connect with nature, why can't they, and does this matter? In this chapter we will argue that it does matter and that engaging with nature is an essential requirement for children's cognitive, emotional and physical development, as well as being good for the planet.

The Nature Problem

A connection to nature and its availability is one that was, until recently, taken for granted. It was assumed that the natural world was present and would be present for future generations. A wake-up call came with the publication of a number of highly influential books and articles, starting in the 1960s with books such as Rachel Carson's *Silent Spring* (1962). This book highlighted the impact of toxins upon wildlife, particularly birds – hence the title, which refers to the loss of bird song. A few years later in 1968 Garrett Hardin's seminal paper 'Tragedy of the commons' drew attention to the impacts of excessive and selfish overconsumption of natural resources. The idea that natural resources are finite and need to be the focus of careful stewardship is one that has gained much credence since and is central to the idea of sustainable development – 'development that meets the needs of the present without compromising the ability of future generations to meet their own needs' (WCED, 1987, p43). The belief that resources need to be preserved for future generations is one that is reiterated by organizations, governments, politicians, environmentalists and

Figure 9.1 Having fun in the mud

Source: Paul Tranter

economists. It is encapsulated in national development strategies, local government documents, business mission statements, educational briefs and is even enshrined in the national planning and environmental legislation of countries such as New Zealand (New Zealand Government, 1991).

International conferences, conventions and treaties recognize the importance and the vulnerability of the natural environment. The United Nations International Year of Biodiversity was launched on 11 January 2010. In promoting this initiative, the UN drew attention to the more than 8400 species currently threatened with extinction, including 1200 bird species. According to the World Conservation Union (IUCN, 2010), one third of the world's amphibian species, one quarter of its mammal species and one eighth of all bird species are endangered (IUCN, 2008). The emphasis on future generations further recognizes the particular impacts of resource overuse upon the world that today's children will grow up to inherit. Will future generations be able to hear the dawn chorus? Not only is there recognition of the damage to the planet, but that in losing nature we also lose something valuable to our own emotional and physical well-being. E. O. Wilson, the famous Harvard-trained entomologist, first used the term biophilia to encapsulate this need, which he defined as: 'The connection that human beings subconsciously seek and need with the rest of life' (Wilson, 1984, p2). It is a connection that is especially important for children for two primary reasons. First, it is important for children's own health and well-being; second, it is important for the well-being of the planet. If children themselves grow up disconnected from nature, how can they value it and act as good stewards for the environment? However, in this regard care has to be taken that children are not seen as responsible for rectifying the follies of their parents, grandparents and forebears and others whose actions have brought about the despoliation of the environment.

If children grow up disconnected from nature, how can they value it?

Why Nature Particularly Matters for Children

Every once in a while a book is published that presents a real challenge to society and society's penchant for 'business as usual'. One such book is Louv's (2008) *Last Child in the Woods: Saving Our Children from Nature-Deficit Disorder*. This disorder is described as follows:

> Nature-deficit disorder is not an official diagnosis but a way of viewing the problem, and describes the human costs of alienation from nature, among them: diminished use of the senses, attention difficulties, and higher rates of physical and emotional illnesses. The disorder can be detected in individuals, families, and communities. (Louv, 2008, p36)

As an indication of the level of their concern about this problem, Louv and others have set up the Children and Nature Network, whose primary aim is 'to help shape a society in which the public once again considers it to be normal and expected for children to be outside and playing in natural areas' (Charles and Louv, 2009, p21). Childhood experiences have been shown to be key determinants of adult interest in nature.

In their study of 'nature and the life course', Wells and Lekies (2006) looked at a raft of studies examining links between childhood experiences and adult environmentalism, and themselves interviewed 2000 adults living in urban US. They found that participants who had experiences with what they called 'wild nature' through hiking, camping, hunting, fishing, walking and play exhibited more positive environmental behaviours as adults. Interestingly, their study, in contrast to other studies, found that for their participants, engaging in formal activities such as environmental education programmes was not a significant predictor of environmental attitudes or behaviours. Experience of nature can take a number of forms. Three are identified here: direct, indirect and observed:

1 *Direct contact:* free interaction – play (e.g. hunting for tadpoles with friends in the local pond or ditch (see Figure 9.2); mediated interaction – work, education, club/group outing (e.g. catching tadpoles and identifying the species in a nature reserve during a school outing).
2 *Indirect contact:* media, internet, books, stories (watching a film on frog development or being told a story by grandma about how she used to catch tadpoles).
3 *Observed without contact:* from car, plane, train, bus (looking at a pond from the car window).

Of these, direct experience is most important for a number of reasons. Children may learn about the life cycle of the frog from TV; but the knowledge is transitory and superficial compared to that gained from actively exploring

Figure 9.2 Catching tadpoles using adapted milk containers

Source: Claire Freeman

where frogs live and seeing the spawn, tadpoles and half-grown frogs in the pond, or bringing them home to rear. In classrooms across the world, watching the tadpoles develop in the fish tank is one of many children's most intimate nature-based school experiences.

Direct contact is essential for 'knowing' nature, and 'knowing' nature is the key to enjoying and conserving it. Lack of such experience can render nature scary, alien and to be avoided. We cannot expect children to be realistic about either the joys or the risks involved in nature if they haven't had any experience. For example, many children, especially those living in cities in hot climates, can encounter snakes when going about their daily lives: playing in the garden, school grounds, by the river or going about daily activities such as attending to the wood pile or bringing in the washing. They need to be able to be realistic and knowledgeable about snakes as they can't and shouldn't avoid all outdoor spaces just because snakes are present. Children cannot meaningfully understand or assess the risk of snakes if:

Lack of direct contact can render nature scary, alien

- They have never experienced snakes.
- They don't know that in most cases the snake's first reaction is to flee from people.
- They don't know that most snakes are not poisonous.
- They don't understand that snakes play an important ecological role – notably, in keeping rodent populations down, which can help to reduce predation on crops and the number of rodents that may harbour diseases injurious to human health.[1]

However, while children need to be taught not to automatically fear snakes, they also need to be taught snake-appropriate behaviour, such as lifting sheets of corrugated iron or wood left in the garden carefully or with a stick in case snakes are under them; wearing closed in shoes in overgrown places and natural bush; and retreating quietly when encountering snakes. They should also be able to identify locally common dangerous snakes. Louv's (2008) worry that nature is being seen as the 'bogeyman' to be feared and avoided is becoming an unfortunate reality; yet few natural experiences are actually dangerous. Most are benign and fascinating, whether it is an encounter with a centipede on the doorstep, a possum or raccoon raiding fruit trees at night, catching crabs, or watching a Children's python snake basking in the sun.[2]

The loss of positive natural experiences is known as 'environmental amnesia'. Thus, if a child's only experience of a river is of the river in a degraded state with no fish or birds, then its degraded state becomes the only experience children have of that river, and they will not expect to see fish and birds. Furthermore, if over each generation the degradation increases, each generation will come to accept that degraded condition as the norm (Kahn and Kellert, 2002, pxiii). This situation is evident for many children growing up in urban areas who may have no experience of a green city and whose parents and grandparents may have no such experience and, thus, no expectations of what could be. Children can be taken to places to see and experience 'better' natural environments; but infusing these distant experiences into and making them relevant to their own life worlds is a major undertaking and cannot compensate for the lack of natural experiences at home (Kellert, 2002).

Health and Well-Being

Contact with nature is necessary for our well-being. Orr (2002) goes so far as to state that it is an essential part of what makes us human. Studies – primarily in the medical field – support the notion that nature is good for us. To date, most of the studies have been carried out on adults, where findings consistently show positive health benefits through contact with nature. A number of studies reported in a paper by Maller et al (2005) show that access to natural views helps patients to recover from operations faster and need fewer painkillers; helps to restore physical energy after stress; reduces stress and other symptoms in prisoners; aids cognitive performance; promotes positive moods; and enhances job satisfaction. Natural environments, they conclude, 'can be seen as one of our most vital health resources' (Maller et al, 2005, p52). Similar benefits for children can be expected. Particular benefits have been noted for children having contact with nature for children suffering from attention-deficit/hyperactivity disorder (ADHD) (Louv, 2008, p107), children experiencing difficulties in their lives (Maller, 2009) and for child hospital patients (Whitehouse et al, 2001). Nature is also good for families: it provides opportunities for families to come together, often in a relaxed setting, and to do activities together. These needn't be 'special' activities, but can include gardening, going for a walk, feeding the ducks or walking the dog (see Figure 9.3).

If contact with nature is shown to have positive emotional and physical benefits, then the reverse is also likely. If we consider human history, it is only during recent times that so many have lived a life divorced from the cadences and rhythms of the natural world. Statistics on the mental and physical health

Figure 9.3 Gardening provides contact with nature, enjoyed across all age groups

Source: Yolanda van Heezik

of children and young people indicate that something is amiss. A 1999 study of children's mental health in the UK, followed up in 2004 and 2007 asserts: 'Among children aged 5–15 years, 5 per cent had clinically significant conduct disorders; 4 per cent were assessed as having emotional disorders – anxiety and depression – and 1 per cent were rated as hyperactive, an overall rate of 10 per cent with some having more than one disorder (Meltzer et al, 1999, p4). Worrying statistics have also been emerging for children's physical health. A 2007 study of US children by Skelton et al (2009) found that 'Rates of severe childhood obesity have tripled in the last 25 years, with significant differences by race, gender, and poverty.' Two studies by a US research team who looked at a total of 7800 children observed that living in areas with green space has a long-term positive effect on children's weight and, thus, health (Bell et al, 2008). The link between deprived environments, poverty and race has been well established. Chapter 3 identified a link between housing and poverty. Experience suggests that areas of housing deprivation with higher levels of poverty and ill health in cities are also areas least likely to have accessible natural green space for children to use, including gardens. Where green space is available, it often takes the form of swathes of mown grass rather than diverse natural environments with rivers, trees, long grass and populated by insects, birds and mammals. If access to nature can be medically proven to ease illness and promote good health (Kaplan and Kaplan, 1989), then the reverse is also likely: lack of access increases the prevalence of illness.

Access to nature can be medically proven to ease illness and promote good health

Wild Experiences: Free Versus Constructed Access

The greatest change in children's play associated with city life during the last generation has been the reduction in children's free access to outdoor play, including play in wild places. Paralleling this change has been a growing emphasis on the importance of the environment and looking after it (recycling, saving water, reducing car use), especially as part of the school curriculum. These two trends are contradictory, for while children are being given (indeed, are bombarded by) environmental messages, they are simultaneously experiencing increasing difficulty in directly engaging with natural environments that they are told can be dangerous and lacking in value (as they are removed and built upon).

Children's play is becoming more and more supervised – if not controlled – by adults. It increasingly takes place in controlled settings with demarcated boundaries, walls, fences and doors, and involves clearly defined activities – for instance, dance or football – usually in a controlled environment with regulated temperatures, water supply, ground material and distribution of objects in set places. By contrast, in natural environments, the child experiences:

- changes in weather (air temperature, wind, precipitation);
- changes relating to the presence or absence of water (the presence of water bodies and the different vegetation types associated with high and low rainfall climates);

- seasonal changes (trees and shrubs in leaf or not, the presence of flowers or fruits, temperature-related activities such as summer swimming or winter ice-skating);
- structural changes (height, gradient, depth);
- changes in types of materials (soft or hard ground, flexible or rigid plants, tough or fragile, long lasting or temporary);
- changes in distribution and layout (natural and formal plantings, plentiful or scarce, dense and scattered, random or patterned).

While trips to the countryside or a nature reserve can provide opportunities to experience some of these changes, unless there is an opportunity for repeat observations and experiences, children's environmental perceptions will be those of a static environment. Only through repeated exposure over time can children gain direct experience and real understanding of the processes of natural change.

Play spaces for children are becoming increasingly denatured: natural play features such as trees and earth mounds are replaced by sterile, manufactured, garishly coloured, safety-conforming tubular steel structures. Trees are removed as their branches can fall and fallen leaves can be slippery; hedges give way to steel fences to enhance visibility; ponds are drained to prevent drowning; duck feeding is banned; logs are removed as they rot and hide insects; long grass is mown; and safety mats cover the ground. Some of these restrictions, as the Audubon Society's leaflet *Please Don't Feed the Waterfowl* explains, are intended for the benefit of nature (Audubon Society of Portland, undated). But if easily accessible wildlife experiences such as feeding the ducks are removed, what alternatives are available? Children, starting as toddlers, love to feed the ducks: for many it is their first and possibly their only way of directly interacting with a wild creature. Contact with nature for children not only needs to be direct, but fun and spontaneous (see Figure 9.4) – hence the importance of duck feeding; catching tadpoles; playing with mud, fish or crabs; skimming stones; jumping in the waves; jumping on stepping stones; building dams in rivers; climbing trees; kicking autumnal leaves; the inevitable duck chasing by younger children; and a myriad of other activities.

The natural world that children meet in cities is one that is robust and adapted to people

The natural world that children are most likely to meet in cities is one that is robust and adapted to people. Few rare and fragile species live in cities; mostly it is populated by the hardy, the common and those species that can coexist with people, that can adapt to noise and disturbance, and that are opportunistic, taking advantage of the food and shelter that the city offers. For children, this hardiness is important: it means that the birds, creatures and plants that they are most likely to encounter can survive disturbance; the tree can withstand climbing; the birds are accustomed enough to people to use a garden feeder; the lizards will carry on sunbathing on walls while the family goes about its business; and flowers can be picked as more will take their place. The behaviours appropriate to special natural places such as nature reserves need to be learned; but for the most part, the city's natural world can be seen, touched, listened to and frequently manipulated and rearranged without too much harm – characteristics that to children make urban nature especially appealing. Access to these experiences is a major issue for many city children.

Figure 9.4 Even formal parks provide playful contact with nature: Watching the fish and ducks in Edogawa Japanese Gardens (Gosford, Australia)

Source: Paul Tranter

Declining Access

For children's free outdoor play, the local neighbourhood is the most important natural play space. A detailed study of changing relationships with nature across the generations in England reveals some disturbing trends (England Marketing, 2009). The study, involving 1150 adults and 502 children, particularly looked at play and natural experiences. Key results are as follows:

- Children spend less time playing in natural places, such as woodlands, the rural countryside and heaths, than they did in previous generations. Less than 10 per cent play in such places compared to 40 per cent of adults when they were young.
- The most popular place for children to play is in their home, while for adults it was outdoors in local streets: 62 per cent of children said that they played at home indoors more than any other place; 42 per cent of adults said they played outdoors in local streets more than in any other place.
- Three-quarters of adults claimed to have had a patch of nature near their homes and over half went there at least once or twice a week; 64 per cent of children reckon that they have a patch of nature near their homes, but less than one quarter go there once or twice a week.
- In the past, favourite places to play were in the streets, near home (29 per cent), indoors (16 per cent) and in some natural places (15 per cent), whereas today children like playing indoors best (41 per cent) and, to a lesser extent, in the garden (17 per cent).
- The majority of children (over 70 per cent) say that they are supervised wherever they play, although only 52 per cent are supervised in the garden and 31 per cent in the streets near their homes. This rises to over 80 per cent in natural places.
- Parents would like their children to be able to play in natural spaces unsupervised (85 per cent).
- Children would like more freedom to play outside (81 per cent).

Another study which analysed children's nature experiences in 16 countries as varied as Morocco and France found similar findings, suggesting that the UK study findings are indicative of wider international trends (Singer et al, 2009). They are further confirmed in a raft of other studies on children's declining outdoor play and exploration (Freeman, 1995; Valentine and McKendrick, 1997; Valentine, 2004; Freeman, et al, 2007; see also Chapter 10).

In Finland children routinely move independently around the city

Fortunately, the decline has been resisted in some places. Studies in Finland show that children routinely move independently around the city and had substantially more freedom than children in comparable UK and German environments (Kytta 1997; Korpela et al, 2002), with children in both Finnish and Belarus urban settings reporting high rates of access to environmental features and activities (Kytta, 2002, 2004). Where children have access to natural landscapes (Fjortoft and Sageie, 2000) and greater independent mobility (Risotto and Tonnucci, 2002; Prezza et al, 2005), they not only engage with nature more, but exhibit better environmental cognition (Wells, 2000).

A number of factors militate against children experiencing these benefits (except in some nations such as Scandinavia). The first is the general loss of independent mobility relating to limitations placed by parents on children. The second is declining tolerance of children's activities. 'Keep off the grass' and 'no ball games' signs pepper the urban environment. The third is the loss of access to natural spaces in cities where children can play. These places may have been built upon; but more usually they are places where access is no longer permitted or where activities which involve experiencing and manipulating the environment, such as tree climbing and lighting fires, have been banned. The fourth is a growing emphasis on indoor play.

'Keep off the grass' signs pepper the urban environment

In his study *No Fear: Growing Up in a Risk Averse Society*, Gill (2007) refers to some of the more extreme cases in this regard: the arrest of three 12-year-olds for climbing a cherry tree in the West Midlands, UK, and a three-year-old and his friends cautioned for 'causing a disturbance' while playing football. A whole raft of activities are increasingly seen as undesirable and some legislated against: activities such as playing on dunes (causes erosion), damming rivers, lighting fires, building structures (dens, tree houses) damaging trees (climbing them), hunting and catching animals, and removing them from their natural environment (taking tadpoles home), and cycling on waste tips. The restrictions being placed on these activities result from growing fears which take a variety of forms: fear that children can hurt themselves; fear that property owners can be held responsible for any accidents that may occur; health fears around diseases that could be contracted from 'dirty' environments; fears of crime that children may either cause (vandalism) or be victims of (abduction). In 2000, the BBC reported on a study by Sarah Thompson that found schools had banned conkers, a traditional seasonal game played by UK children where horse chestnuts suspended on a string are bashed against each other until one breaks. One school was reported to have purchased safety goggles for children to wear while playing the game. The report resulted in a flurry of responses about the 'nanny state' and its impact upon children (HSE, 2007).

Real dangers do exist in natural environments; but only by exposure can children acquire the skills necessary to judge them: for example, which branches on which trees can hold their weight; which plants sting and should be avoided; and the difference between frogs and toads (toads can emit noxious secretions from their skin). When walking in Australia, children need to learn in which places magpies nest and how wide a berth to give to them. Australian magpies can be particularly aggressive during nesting, dive-bombing people who come too close. Warning signs are erected by some councils where aggressive magpies are known to lurk. Many cyclists have spikes sticking out of their helmets to deter attackers. On one occasion, one of the authors was driven out of a picnic spot by a magpie that continually attacked a child wearing a blue t-shirt and gave him a nasty bruise on his forehead. All other children were ignored. A common response to the 'dangers' associated with natural areas, or even the outdoors, in general, is to restrict children to their immediate surrounds. Especially common are children whose only play space is their home and its garden, or the space immediately in front of their home, although even this space is becoming less available due to a number of trends in house construction:

- bigger houses on smaller plots, leaving little outdoor space (the more ostentatious ones have become known as McMansions);
- new housing estates that prioritize indoor over outdoor space, and like the McMansions, leave little outdoor play around the house – such houses can

cover up to 60 per cent more of the plot/garden than traditionally built smaller homes (Hall, 2008) (ironically, these estates are invariably classified as 'family-friendly' environments);
- conversion of front yards/gardens into car parking spaces;
- infill development (erecting additional buildings, flats and homes on what was previously a single-home site);
- developments that lack footpaths/sidewalks (access is presumed to be by car).

The focus on the house in part reflects the fact that children are spending much greater proportions of their time indoors and are engaged with manufactured play items. In a much-quoted study, researchers found that young children clearly had tremendous capacity for learning about creatures. Unfortunately, the creatures they were considerably better at identifying were synthetic Pokémon creatures, rather than the real creatures, such as oak trees or badgers (Balmford et al, 2002). Unless children access the outdoors and encounter nature, this superior knowledge of imaginary creatures will persist. Yet, as the next section shows, most cities are, nonetheless, still full of wildlife and provide many opportunities for children to have contact with real creatures.

Nature Around Us

Cities are mythologized as places devoid of nature

Cities are mythologized as places devoid of nature, as a rapacious sprawling spider whose tentacles are ever increasing and gobbling up the countryside, depleting its natural treasures. As ever with such images, there is some element of reality; but the overall image is misleading. Wildlife is present in most cities, but is often taken for granted, as with pigeons, or overlooked, as with most insect species (see Figure 9.5). The countryside is often far from natural: it is intensively farmed and managed; pesticides abound; there is little food or shelter for wildlife; native vegetation is removed to make way for more commercially or aesthetically desirable plants; and the edge of the city is frequently home to farm-type smallholdings and businesses that need space and may be noisy (e.g. dog kennels, machinery storage and quarries). Cities have increasingly become refuges for species, some no longer able to survive in the countryside or who have opted for the urban life. Several of the better-known species include coyotes moving into US suburbs (a place where raccoons have long since been at home), foxes in European cities, monkeys that thrive in many Asian cities, and flying foxes, brush turkeys and bandicoots in Australia. Some, like bats, have become tourist attractions. In Austin, Texas (US), there are tours to see the 1.5 million Mexican free-tail bats that roost on the Congress Street Bridge, while in Brisbane, Australia, there are tours along the river to Indooroopilly Island not far from the central business district (CBD), where hundreds of thousands of flying foxes camp. Brisbane is a city that is replete with wildlife: in an average day a child living in the CBD could easily see several species of parrots (including cockatoos) flying around, encounter water dragons by the river, see large lizards on the school playing fields and skinks on walls, meet a bush turkey walking along the footpath, and at night listen to the flying foxes chitter as they fly overhead, be kept awake by the 'kuk kuk' noise of the now ubiquitous house geckos that lurk behind pictures, and encounter the 'dreadful' cane toad (exotic and invasive, they are poisonous to people and

Figure 9.5 Children value the everyday contact with nature that common, usually overlooked garden species such as this stick insect provide

Source: Claire Freeman

native wildlife) if they wander outside on a rainy evening. Little wonder, then, that Tim Low (2002, p3), himself a Brisbane resident, writes:

> Nature is sold to us as something separate that lives far away from us in wild places, when really it's all around us engaging us more than we guess. The wilderness begins right here where we live.

Is Brisbane exceptional? The answer is perhaps. Its wildlife is large, colourful, noisy and hard to miss. Not all wildlife is so easily observed or encountered; but in most parts of most cities, it is, nevertheless, present, but under threat, as wasteland is tidied and built upon and the urge to maximize profit from urban land marches inexorably onwards.

Green Cities

Cities tend to be immensely rich and varied in green space. Depending upon the size, culture and geographic location of the built-up area, a child could easily encounter in their daily lives any of the green spaces summarized in Table 9.1. The presence of green space and wildness in cities has been undervalued and overlooked. Many schools provide 'nature trips' to places where children can encounter nature. For children from densely urbanized inner-city environments, such trips may provide opportunities to meet nature that is otherwise unavailable; but the majority will live in places that are not devoid of natural experiences. One of the early pioneers in recognizing this natural wealth was Ann Whinstone Spirn (1984), who described the city as a granite garden utterly dependent upon natural processes. In a similar vein, Michael Hough (1995) talked of how the city is divided into pedigree and non-pedigree landscapes. Pedigree landscapes are those with formal landscaping, assiduously maintained by human management, unlike the non-pedigree, where nature maintains itself in estuaries, wastelands and unmanaged spaces. Children do not always do well in the pedigree landscape, but can thrive in the non-pedigree, as the early work of Hart (1978) demonstrated. In his study of children in Vermont, the children showed Hart their special places, invariably non-pedigree in form, where they could play, make a 'mess', construct things and use the natural environment as their playground.

Cities are moving forward and increasingly recognizing and supporting nature in the city, as is indicated in the growth of urban wildlife groups, the relaxation of landscape management to encourage regeneration, the protection of natural sites and the development of urban biodiversity strategies. In schools there has been a strong move through organizations such as Learning through Landscapes (UK), Enviroschools (New Zealand), Ecoschools.US (US) and the Ecoschool programme (Japan and Sweden) in creating more natural school grounds with natural landscaping, including nature gardens, ponds and vegetable gardens (see Chapter 4). In parks, regeneration is being encouraged underneath mature trees; some lawns are being allowed to grow, creating

Table 9.1 Urban green spaces

Green space categories	Type of spaces
Public institutions	Hospital, health service, church, grounds; school playing fields; cemeteries/burial places; government buildings
Private space	Gardens, private or communal; courtyards
Landscaping	Street trees, flower beds, landscaped areas/city squares
Sports/play	Sports fields, playgrounds
Industrial	Landscaping around businesses, naturally regenerating wasteland
Transportation spaces	Road verges, landscaped traffic islands, railway sidings
Natural features	Overgrown walls, bird and bat roosts, individual trees, hedges
Aquatic	Lakes, rivers, ponds, canals, ditches, sewage ponds
Parks	Botanic gardens, parklands, playgrounds and sports parks
Nature areas	Woodlands, wetlands, grasslands, reserves and protected habitats
Productive land and farms	Fields, market gardens, orchards, grazing paddocks, city farms, communal gardens

wildflower meadows, and native plants and habitats are replacing exotic ones. Wonderful examples of this can be seen in Pinaroo Cemetery in Perth, Australia, where lawns are grazed by wild kangaroos. In London, the 42ha London Wetland Centre, reputedly the best urban site in Europe to watch wildlife and an international award-winning visitor attraction and Site of Special Scientific Interest (SSSI) (see Figure 9.6), has been created from disused concrete reservoirs that were part of London's water supply system. The centre charges entry fees which can limit access. Yet, it does provide children, primarily though its schools' programme, with a fabulous opportunity to experience wild and rare birds without leaving the central city.

While the importance of places such as the London Wetlands Trust site are self-evident, how easy is it for a child to access wildlife in their daily lives? A random, admittedly somewhat unscientific, search using Google Earth satellite photographs was undertaken to explore this question. Putting in a fictitious address – namely, 16 Church Street/Road, or its closest equivalent, depending upon language and context – Table 9.2 was constructed to show what type of green spaces were available in easy walking distance of the home of a child living at that fictitious address. What the sample shows is that for most children, there would be access to a range of green spaces of varying degrees of

Figure 9.6 Child using the interactive outdoor displays at the London Wetland Centre

Source: Claire Freeman

naturalness. It shows that all would have access to some private green space, a park or some sort of institutional grounds, and most would have easy access to vegetated wasteland, groups of trees and some sort of natural area. While Tokyo, Madrid and Paris have few easily available natural areas outside parks, the majority of addresses in Table 9.2, by contrast, reveal access to a range of varied spaces. What is particularly interesting are the range of places available to children in high-density housing, such as in Hong Kong, Rotterdam and Warsaw. For children growing up in 'Church Street/Road' in cities internationally, the physical environment, for the most part, still offers the opportunity for natural experiences of the sort that shaped the childhood of Bellamy, Attenborough and Goodall in London. The question is: are children today able to engage with natural places in their home environment? In order to ensure that they are, some changes will need to be made and some positive policy initiatives implemented.

Table 9.2 Types of green spaces easily accessible to children living at a random sample of addresses, based on a search of 16 Church Street/Road or its closest equivalent

	Type of area	Private gardens	Sports fields	Groups of trees	Park	Play area	Field	School/church	Nature/natural area	Vegetated wasteland	Countryside	Beach/river/lake
16 Church Road, Seaforth Liverpool, Merseyside	City: industrial terraces	✓	✓	✓	✓	✓	✓	✓		✓		✓
16 Church Road, Melbourne Australia	City outer suburb: houses	✓		✓	✓				✓	✓		
Church Street, Shau Kei Wan, Hong Kong	City: apartment/ tower blocks	Cy	✓	✓	✓	✓		✓	✓	✓	✓	
Church Road, Kahnawake, Montreal, Canada	City suburb: house	✓	✓	✓	✓	✓	✓	✓	✓	✓	✓	✓
16 rue de l'Eglise, 75015 Paris, France	City: apartments	✓		✓	✓			✓				✓
Kirchstraße 16, 13158 Berlin, Germany	City outer suburb: house	✓	✓	✓	✓			✓	✓	✓		
Kyu yamati Dori, Tokyo	City: commerce – apartments	Vs	✓	✓	✓			✓				
Calle de la Iglesia, 28019 Madrid, Spain	City: apartments	Vs			✓	✓	✓	✓		✓		✓
6254 Church Street, Los Angeles, California	City outer suburb: houses	✓	✓	✓	✓	✓		✓	✓	✓		
Droga kościół, Warsaw	City: apartments	Cy	✓	✓	✓	✓			✓	✓		✓
Kerkstraat 16, 3054 Rotterdam, The Netherlands	City: apartments	Cy	✓	✓	✓	✓	✓	✓	✓	✓	✓	✓

Notes: Cy = courtyard; Vs = very small. Note that the green spaces were identified using aerial photography and are indicative only; it is possible that other green spaces were present but not picked up due to identification difficulties.

Re-engaging with the 'Wild'

Any re-engagement policy will need to have two strands. The first is that children's ability to experience their surroundings must be encouraged and reinstated. The second is the need to ensure that there are places where children can experience wildness. The first accords with one of the key themes of this book: reinstating children's right to roam, play and be outdoors. Sue Palmer's (2006) book *Toxic Childhood* has a parental checklist for 'detoxing the great outdoors'. The most pertinent criteria is to 'resist irrational fears, and balance worries about letting your child play outside with the knowledge of how important outdoor play is'. The message equally applies to landowners, educators, agencies and others with overzealous worries over litigation and possible dangers to children's health. Natural spaces in a variety of forms need to be recognized as potentially valuable play space. The urge to automatically 'tidy' should be resisted, recognizing that some indication needs to be given that a place is important, its 'messiness' and 'untidiness' intentional and not just evidence of managerial neglect (see also Chapter 4). Oliver Gilbert (1989), the Sheffield (UK)-based author of *The Ecology of Urban Habitats*, supported mowing the edges of wild spaces to indicate that the space was being managed, and to complete the job a sign would be erected stating that the 'messy' overgrown area behind the mown strip was natural habitat. A similar approach could be applied to natural play spaces: erect a sign saying 'This area supports children's play in natural places.' There will also need to be acceptance of children's activities. While it may be necessary to discourage children from catching tadpoles in ponds that are home to rare and endangered aquatic wildlife, in most cases city ponds are populated by common species whose overall populations will not be harmed by some catching activities.

There is a need to ensure that there are places where children can experience wildness

Natural spaces will need to be provided for children living in the most naturally deprived environments; in Table 9.2 these would be in central Madrid, Rome and Tokyo. Where natural environments are created, they should be clearly indicated as open to visitors, at no or minimal cost, and should be part of the school curriculum so that all children can experience them and hopefully then make repeat visits on their own, with friends or family. These sites do not necessarily need to be completely wild environments, but must be predominantly vegetated, robust environments that can also be used for play. Two such environments that are becoming increasingly common in larger cities are green roofs and city gardens. By 2006, Chicago had over 200 green roofs and city gardens; by 2009, the New York City Department of Parks and Recreation listed over 600 member gardens. An increasing number of children's gardens located in the inner city are being developed which provide accessible garden and play space for local communities and can fill an important gap in the open space resource for inner-city children. In the long run, visits to local places where children can develop a lasting relationship will be more beneficial than one-off visits to more exotic and impressive environments.

Notes

1 Snakes are not normally aggressive. The Florida Museum states: 'Regardless of what some people say, Florida snakes are not aggressive, and unless they are cornered, most will flee when they see you' (see www.flmnh.ufl.edu/herpetology/fl-guide/gettingalong.htm). How venomous are snakes? Out of the 2800 kinds of snakes found in Australia, only 270 types are venomous (see www.australiazoo.com.au). For snakes' ecological role, see Ostfeld and Holt (2004).
2 The Children's python (*Antaresia childreni*) is a small non-venomous python that is found throughout much of the north and north-east of Australia. The Children's python is often kept as a pet as it is small, easy to look after and non-venomous. Since it is a common house pet, many mistakenly believe that the name Children's python refers to its ease of care. Actually, its name comes from its dedicatee, John George Children (see www.wilddownunder.com/wildlife).

References

Attenborough, D. (2009) 'Interview', *The Big Issue, Scotland*, 25 June 2009
Audubon Society of Portland (undated) *Living with Wildlife: Please Don't Feed the Wildfowl*, Wildlife Care Center, Portland, OR
Balmford, L. Clegg, T., Coulson, T. and Taylor, J. (2002) 'Why conservationists should heed Pokémon', *Science*, vol 295, p2367
Bell, J. F., Wilson, J. S. and Liu, G. C. (2008). 'Neighbourhood greenness and 2-year changes in body mass index of children and youth', *American Journal of Preventive Medicine*, vol 35, no 6, pp547–553
Bellamy, D. (2002) *A Natural Life by David Bellamy*, Arrow, London
Carson, R. (1962) *Silent Spring*, Houghton Mifflin Company, New York, NY
Charles, C. and Louv, R. (2009) *Children's Nature Deficit: What We Know – and Don't Know*, Children and Nature Network, www.childrenandnature.org/
England Marketing (2009) *Report to Natural England on Childhood and Nature: A Survey on Changing Relationships with Nature across Generations*, www.englandmarketing.co.uk
Fjortoft, I. and Sageie, J. (2000) 'The natural environment as a playground for children: Landscape description and analyses of a natural playscape', *Landscape and Urban Planning*, vol 48, pp83–97
Freeman, C. (1995) 'The changing nature of children's environmental experience: The shrinking realm of outdoor play', *International Journal of Environmental Education and Information*, July–September, vol 14, no 3, pp259–281
Freeman, C., Quigg, R., Vass, E. and Broad, M. (2007) *The Changing Geographies of Children's Lives: A Study of How Children in Dunedin Use Their Environment – Report on Research Findings*, Geography Department, University of Otago, New Zealand
Gilbert, O. L. (1989) *The Ecology of Urban Habitats*, Chapman & Hall, London
Gill, T. (2007) *No Fear: Growing Up in a Risk Averse Society*, Calouste Gulbenkian Foundation, London
Goodall, J. (1996) *The Magic I Knew as a Child,* http://ecopsychology.athabascau.ca/0996/index.html, accessed February 2010
Hall, T. (2008) 'Where have all the gardens gone?', *Australian Planner*, vol 45, no 1, pp30–37
Hardin, G. (1968) 'The tragedy of the commons', *Science*, vol 162, pp1243–1248
Hart, R. (1978) *Children's Experience of Place: A Developmental Study*, Irvington Publishers Halstead/Wiley Press, New York, NY

HSE (Health and Safety Executive) (2007) *Great Health and Safety Myths*, www.hse.gov.uk/myth/index.htm

Hough, M. (1995) *City Form and Natural Processes*, Routledge, New York, NY

IUCN (World Conservation Union) (2008) *State of the World's Species Factsheet*, http://cmsdata.iucn.org/downloads/state_of_the_world_s_species_factsheet_en.pdf

IUCN (2010) Red List: State of the World's Species, http://cms.iucn.org/iyb/, accessed 2010

Kahn, P. H. and Kellert, N. M (2002) 'Introduction', in P. H. Kahn and S. R. Kellert (eds) *Children and Nature: Psychological, Sociocultural and Evolutionary Investigations*, MIT Press, Cambridge, MA

Kaplan, R. and Kaplan, S. (1989) *The Experience of Nature: A Psychological Perspective*, Cambridge University Press, New York, NY

Kellert, S. (1997) *Building for Life: Designing and Understanding the Human–Nature Connection*, Island Press, Washington, DC

Kellert, S. (2002) 'Experiencing nature: Affective cognitive and evaluative development in children', in P. H. Kahn and S. R. Kellert (eds) *Children and Nature: Psychological, Sociocultural and Evolutionary Investigations*, MIT Press, Cambridge, MA

Korpela, K., Kytta, M. and Hartig, T. (2002) 'Restorative experience, self-regulation and children's place preference', *Journal of Environmental Psychology*, vol 22, pp387–398

Kytta, M. (1997) 'Children's independent mobility in urban, small town and rural environments', in R. Camstra (ed) *Growing Up in a Changing Urban Landscape*, Royal van Gorcum, Assen, pp41–52

Kytta, M (2002) 'Affordances of children's environments in the context of cities, small towns, suburbs and rural villages in Finland and Belarus', *Journal of Environmental Psychology*, vol 22, no 1, pp109–123

Kytta, M. (2004) 'The extent of children's independent mobility and the number of actualized affordances as criteria for child-friendly environments', *Journal of Environmental Psychology*, vol 24, pp179–198

Louv, R. (2008) *Last Child in the Woods: Saving our Children from Nature-Deficit Disorder*, Algonquin Books, Carolina

Low, T. (2002) *The New Nature Winners and Losers in Wild Australia*, Penguin Australia, Australia

Maller, C. J. (2009) 'Promoting children's mental, emotional and social health through contact with nature: A model', *Health Education*, vol 109, no 6, pp522–543

Maller, C., Townsend, M., Pryor, A., Brown, P. and St Leger, L. (2005) 'Healthy nature healthy people: "Contact with nature" as an upstream health promotion intervention for populations', *Health Promotion International*, vol 21, no 1, pp45–54

Meltzer, H., Gatward, R. with Goodman, R. and Ford, T. (1999) *The Mental Health of Children and Adolescents in Great Britain: Summary Report 1999*, Social Survey Division of the Office for National Statistics on Behalf of the Department of Health, the Scottish Health Executive and the National Assembly for Wales Office for National Statistics, UK Government, London

New Zealand Government (1991) *The Resource Management Act*, New Zealand Government, Wellington, New Zealand

Orr, D. W. (2002) 'Political economy and the ecology of childhood', in P. H. Kahn and S. R. Kellert (eds) *Children and Nature: Psychological, Sociocultural and Evolutionary Investigations*, MIT Press, Cambridge, MA

Ostfeld, R. S. and Holt, R. D. (2004) 'Are predators good for your health? Evaluating evidence for top-down regulation of zoonotic disease reservoirs', *Frontiers in Ecology and the Environment*, vol 2, no 1, pp13–20

Palmer, S. (2006) *Toxic Childhood*, Orion Publishers, UK

Prezza, M., Alparone, F., Cristallo, C. and Luigi, S. (2005) 'Parental perception of social risk and of positive potentiality of outdoor autonomy for children: The development of two instruments', *Journal of Environmental Psychology*, vol 25, pp437–453

Rissotto, A. and Tonucci, F. (2002) 'Freedom of movement and environmental knowledge in elementary school children', *Journal of Environmental Psychology*, vol 22, pp65–77

Singer, D. G., Singer, J. L., D'Agostino, H. D. and DeLong. R. (2009) 'Children's pastimes and play in 16 nations: Is free-play declining?', *American Journal of Play*, vol 1, no 3, winter, pp284–312

Skelton, J. A., Cook, S. R., Auinger, P., Klein, J. D. and Barlow, S. E. (2009) 'Prevalence and trends of severe obesity among US children and adolescents', *Academic Pediatrics*, vol 9, no 5, pp322–329

Spirn, A. W. (1984) *The Granite Garden: Urban Nature and Human Design*, Basic Books, New York

Valentine, G. (2004) *Public Space and the Culture of Childhood*, Ashgate, Aldershot, UK

Valentine G. and McKendrick J. (1997) 'Children's outdoor play: Exploring parental concerns about children's safety and the changing nature of childhood', *Geoforum*, vol 28, no 2, pp219–235

WCED (World Commission on Environment and Development) (1987) *Our Common Future*, Oxford University Press, Oxford, UK

Wells, N. M. (2000) 'At home with nature: Effects of "greenness" on children's cognitive functioning, *Environment and Behavior*, vol 32, no 6, pp775–795

Wells, N. and Lekies, K. S. (2006) 'Nature and the life course: Pathways from childhood nature experiences to adult environmentalism', *Children, Youth, Environments*, vol 16, no 1, pp1–25

Whitehouse, S., Varni, J. W., Seid, M., Cooper-Marcus, C., Ensberg, M. J., Jacobs, J. R. and Mehlenbeck, R. S. (2001) 'Evaluating a children's hospital garden environment: Utilisation and customer satisfaction', *Journal of Environmental Psychology*, vol 21, pp301–314

Wilson, E. O. (1984) *Biophilia*, Harvard University Press, Cambridge, MA

Part III

Making a Difference: Creating Positive Environments for Children

10
Accessing Space: Mobility

> A city should be so constructed so that it is safely navigable by any seven-year-old on a bicycle. (Enrique Peñalosa, former mayor of Bogotá, Columbia)

Mobility: Independent or Adult Dependent?

Many parents in modern Western cities marvel at the advantages children have today over children in previous generations. Children today have access to the seemingly limitless world of the internet. They have a variety of 'canned entertainment' at their fingertips (e.g. Xbox, IPod, PlayStation, Game Boy and Wii), so they should never be bored. Many are engaged in stimulating extra-curricular activities after school and at weekends. They have parents who can drive them to these activities, as well as to school, to the cinema or to their friends' homes. Despite these apparent 'advantages' for children today, parents increasingly are asking themselves: 'Are our children really better off than we were as children, or are they missing out on some things?' We can broaden this question by asking whether entire cities are as friendly for children now as in previous generations.

While some children 'benefit' from an apparent ability to widely access the city and its facilities due to their parent's ability to drive them, this is not the case for all children. Cars, and enhanced mobility for some children, also enhance social polarization, as many children do not have access to cars or parents who are able and willing to 'ferry them'. The increased use of private vehicles impacts negatively upon children living in areas where these additional vehicles pass through: their independent mobility may be reduced without any compensating access to private transport.

Do children have 'independent mobility' or 'adult-dependent mobility'?

The chapters in Part II of this book are about spaces for children in cities. We start Part III with a chapter examining children's access to these spaces. The key issue for this chapter is whether children have 'independent mobility' or whether their access is based on 'adult-dependent mobility' or 'car-dependent mobility'. This chapter begins by explaining the value of children's independent mobility (CIM) in comparison to adult-dependent mobility (ADM). It summarizes research on CIM in a range of countries (e.g. the US, the UK, Australia, Germany, Canada and New Zealand), which shows huge contrasts between nations in the levels of children's freedoms, and dramatic reductions in children's freedoms over the last few decades in some countries. Reasons for the international differences and the declines in children's freedoms are outlined. We suggest strategies for increasing children's independent mobility, and conclude with an explanation of how changing our cities so that CIM is enhanced will help to make our cities more liveable for all city residents.

Why Independent Mobility Matters

Does the loss of children's freedom really matter? Should we accept that children's lives are different now, and that adult-organized activities and indoor and virtual activities now replace their independent movement? The freedom of children to independently explore their own environment is of far more value than we may first realize. Children's freedom of movement, in combination with diverse environmental affordances to facilitate play, is fundamentally important in a child-friendly environment (Hart, 1979; Moore, 1986). Not only is independent mobility important for children, it is also of value for their parents and other adults responsible for them, for the wider environment, and for the strength and vitality of local neighbourhood communities. Ultimately, the level of children's independent mobility is an indicator of the success and resilience of a city.

CIM is important for children as it gives them opportunities to explore the world and their own reaction to it through the medium of play (see Figure 10.1). The opportunity to move freely within their own neighbourhood is recognized by children themselves as one of the main positive indicators of an urban environment (Chawla, 2002). It provides them with opportunities for contact

Figure 10.1 Children using the street as a play space

Source: Morgan Loiterton

with nature, particularly if there are any 'wild spaces' accessible to them (see Chapter 9). Such contact with nature is important for the health and well-being of all people, particularly children (Holbrook, 2009).

Freedom to independently explore and to partake in unstructured outdoor play is vital for children's development, not only for their levels of activity (their physical development), but also for their emotional, social and cognitive development (Moore, 1986; Kegerreis, 1993; Fjortoft and Sageie, 2000; Kytta, 2004; Burdette and Whitaker, 2005). Children who do not have this freedom find it difficult to develop a sense of place: a sense of connection with their own neighbourhood and community. As Engwicht (1992, p30) argues: 'robbing children of a sense of place robs them of the very essence of life'.

Children do not gain this sense of place from the back seat of a car: they may see more, but they learn less. The ultimate shielding of children from this sense of place is found in situations where not only are children transported by car, but they are 'entertained' along the way with DVDs played through monitors inside the car. Children in this situation have little sense of place. They have minimal appreciation of where they are, no contact with the local environment or community, and get no developmental experiences that connect them to the real world.

Children do not gain a sense of place from the back seat of a car

When children have lower levels of CIM, traffic danger increases, particularly around the schools and other traffic generators, and fears of 'stranger danger' also increase, as there are fewer eyes on the street. Children learn less about road safety when they are inside cars, and they are also exposed to higher levels of pollution from in-car pollution (International Center for Technology Assessment, 2000). Overall, pollution levels in the city are higher (from the extra traffic involved in ferrying children). Children also miss out on the joy and wonder of taking their own time on the journey home from school (O'Brien, 2003).

As well as being of value for children, CIM is important for parents and adults tasked with looking after children. Parents can spend vast amounts of time transporting and supervising children. Today in Australia, parents spend about twice the time on these tasks that they did a generation ago. The economic resource cost of transporting children is huge. Estimates in the UK of this cost – of transporting children to school, to sport, to their friends houses or even to the local park or playground – suggested an annual figure of between UK£10 billion and UK£20 billion (in 1990 currency) (Hillman et al, 1990).

The effect of the loss of CIM can be appreciated in terms of the number of person trips now made to get children to and from school. A generation or more ago, when a child travelled to and from school, only two person trips were involved: the child going to school unaccompanied, and returning home unaccompanied (walking, cycling or using public transport). Now six person trips are made per day for each child in many households. A parent drives (or accompanies) a child to school (two person trips), then returns home (one person trip). Then, in the afternoon, the parent drives to school (one person trip) and returns home with the child (two person trips). In Melbourne, Australia, 40 per cent of all trips made to take children to school are simple home-to-home chains (Morris et al, 2001). One of the most common reasons given by parents for driving their children to school is time pressure (Mackett, 2002).

Yet, as explained in Box 10.1, getting children to walk to school can save parents time (Tranter, 2010).

There are significant environmental costs associated with reductions in CIM. Large amounts of extra traffic are generated by the current practice of driving children to school (Mackett, 2002). In some US cities, 40 per cent of the morning peak hour is attributed to passenger trips from children being driven to school (Crider, 2003). In Melbourne, trips to take children to school make up 21 per cent of all trips in the morning peak (Morris et al, 2001). This extra traffic contributes significantly to traffic congestion, local air and noise pollution, and carbon dioxide (CO_2) production. In many cities, during school holidays, traffic congestion is noticeably alleviated when school runs are not being made. It is not only the large and growing number of trips to school that contribute to pollution. Passenger trips to take children to a range of

Box 10.1 How walking to school can save parents time

Many parents drive their children to school in an attempt to save time. However, the collective and long-term impacts of driving children to school lead to parents having less time than if they allow their children to walk to school, even if a parent accompanies their child on the walk to school. Parents who use the 'second car' to drive their children to school contribute in complex ways to an environment in which time is increasingly occupied by the various demands of the car. This can only be understood when the feedback effects of parents using cars are considered. Some of these feedback effects are local and immediate, while others are more large scale and long term in their effect.

Parents spend a considerable amount of time at work earning the money to pay for a second car that 'might' save them 30 minutes a day on the journey to school (compared with walking). If they have a large 4WD (or SUV) vehicle, they could be spending more than 2.5 hours per day earning the money just to pay for the direct costs of their second car. If we stopped this argument here, we can see that instead of saving time, parents are, in fact, wasting time. Yet, there are other ways in which driving their child to school robs parents of their time. Because their children don't get exercise walking to school, parents must drive them to organized sport, spending several more hours in their cars each week. Because local streets are too dangerous (because of the traffic generated by parents driving their children around), parents feel that they must also drive their children to the local playground. Because their children don't know other children in the local area, they have to be driven to their friends' houses.

If parents let their children walk to school, even if they walk with their children, this can lead to a saving of time, not just through the time saved by not having to pay for the second car. When children walk to school, they get to know local friends, both on the journey to school and in play activities at their friends' houses, which they are more likely to know about and be able to visit independently. Children who walk to school are more likely to feel a part of the local community, which is also likely to have health benefits, both for adults and children. Children have less need to be driven to sport, as they can play independently in their local neighbourhoods, and are more likely to be allowed to walk or cycle to sporting activities.

By thinking about the journey to school in a more holistic sense, we can see that those parents and children who drive to school have 'less time' and 'less money'. The parents and children who walk to school have not only saved time and money, but they are more likely to be happier, fitter and healthier as well.

destinations represent one of the fastest growing types of urban car travel over the last ten years. In Sydney, the number of serve-passenger trips grew at the rate of 2.8 per cent per annum between 2001 and 2006, much faster than the growth in total driver trips (1.6 per cent), which itself was much higher than the population growth (0.9 per cent) (Shaz and Corpuz, 2008). Parents are now making more trips to take their children to their friends, the park, the playground or to sport. Instead of playing locally, or getting themselves to their friends' houses, children now rely on their parents to drive them.

Another reason why CIM matters relates to its role in helping to generate stronger neighbourhood-based communities. Children are very effective at breaking down the learned reserve between adults, thus contributing to the development of social capital (Weller and Bruegel, 2009). Children playing in and around local streets help to generate the connections between families that contribute to social capital, and health and well-being (Offer and Schneider, 2006). Recent trends in CIM in many nations show that an appreciation of the value of this mobility for children, their parents and the wider environment and community has been overshadowed by the assumed benefits of more car-dependent lifestyles. The following sections outline some international comparisons in CIM and the extent of the decline in CIM over the last few decades in several countries.

Children break down the learned reserve between adults, contributing to social capital

International Comparisons of Children's Independent Mobility

Some of the earliest data on CIM come from England and Germany, where Hillman et al (1990) conducted research for their landmark study *One False Move: A Study of Children's Independent Mobility*. This study reported the results of a study of English schools surveyed in 1971, and the same schools surveyed again in 1990, along with comparisons of schools in similar towns and cities in Germany in 1990. The study design was based on the 'licences' given to children by their parents – for example, licences to travel to school alone, catch buses alone, cross main roads alone, or visit places other than school alone (Hillman et al, 1990). The age at which children are given these licences, or the percentage of children in any group who have them, is an indicator of the level of freedom that children have, or their level of independent mobility. This study showed marked contrasts between the English and the German schools, with the German children having very high levels of freedom compared with the English schools. These differences existed for all of the licences, with the German children having very high levels of freedom from the age of seven upwards.

The research design for the Hillman et al (1990) study has been replicated by other studies in other countries, allowing further international comparisons to be made. Tranter (1996) conducted research in Australian and New Zealand cities, which identified levels of freedom for children that were lower than the levels for German and English school children. For example, for the licence to travel to places other than school alone, German children were almost twice as likely to be given this licence for children aged 9, 10 and 11. Very low numbers of German children were driven to school or to other places compared with English or Australian children.

In other research, international comparisons of levels of walking and cycling for six- to nine-year-old children found that Australian and New Zealand children had less freedom than children in Canada and Sweden, despite the climate being more favourable to these modes in Australia (Timperio et al, 2004). In Canada, even during winter, many children walk to school. Recent Canadian programmes to encourage higher levels of walking to school include Active and Safe Routes to School and the Walking School Bus programme. The walking school bus is a group of children walking with adults to school, picking up children along the way (see Figure 10.2).

Levels of CIM are still high in some European and developed Asian nations. Finland and Japan are two countries where children have very high levels of freedom, despite recent trends towards a reduction in children's autonomy (Horelli, 2001, Thomson, 2009). In Japan, autonomous mobility is encouraged in children from a young age, and a culture of looking out for children and

Figure 10.2 A walking school bus to P. L. Robertson Public School, Ontario (Canada): This school is a part of a pilot Active and Safe Routes to School project across the Halton Region

Source: Jacky Avent

keeping them safe (e.g. on their way to school) has been maintained, even with the growth in car ownership. In Finland, very high numbers of children still walk to school and the majority of them play near their homes without constant adult supervision. Yet, even in these countries, signs of a reduction in children's freedom are emerging.

Trends in Levels of Children's Independent Mobility over Recent Decades

A remarkable decline in the level of children's freedom to explore their own neighbourhood and city independently has occurred over the last few decades in the US, Canada, the UK, Australia and New Zealand, and has also been evident, though probably to a lesser extent, in many European countries.

A remarkable decline in children's freedom to explore on their own has occurred

Hillman et al's (1990) study identified dramatic declines in children's freedoms in English schools between 1971 and 1990, for all licences. This study is particularly significant given that the same five schools were studied using the same research design. While a decrease in children's freedom during this time might be expected, the extent of the decline is alarming. For example, in 1971, 80 per cent of seven- to eight-year-olds were allowed to travel to school independently (by themselves or with other children). By 1990, this had fallen to a mere 9 per cent (Hillman et al, 1990, p42).

Further research in England (O'Brien et al, 2000) shows that the freedom of children (their licences) continued to decline further since 1990, and by 1998, children in English schools were likely to have substantially less freedom than they did during the early 1990s (see Figure 10.3). 2005 data on the mode of travel to school indicates that the trend towards less independent mobility and greater car-based travel is continuing, as is shown in changes in the main mode of travel for primary school children in the UK to get to school (five- to ten-year-olds) (Barker, 2006). While the decline in CIM has occurred for children of all ages, it is most marked for children under the age of 11, with children over this age maintaining much of their independent mobility from the 1990s (Barker, 2006). Studies of changes in CIM in several countries indicate that this decline in children's freedom identified in English towns and cities is also evident in other Western industrialized nations, including Canada, Denmark, Italy, Sweden, Australia and New Zealand (Collins and Kearns, 2001; Fotel and Thomsen, 2004; Rissotto and Tonucci, 2002; Sandqvist, 2002; Gilbert and O'Brien, 2005; Tranter, 2006).

As CIM decreased, adult-dependent mobility increased, so that children are now much more likely to be escorted by their parents, particularly by car. Gilbert and O'Brien (2005, p10) note that in Toronto, Canada, the increase in car trips for children aged 11 to 15 was 83 per cent between 1986 and 2001, in comparison to an increase of only 11 per cent in car use by adults. The declines in CIM in Australia have also been marked over the last few decades. Between 1971 and 2003 there has been a threefold increase in the number of children aged five to nine being driven to school, while the number of children walking to school has halved during this time (see Figure 10.4). Alarmingly, similar trends are evident for children aged 10 to 14.

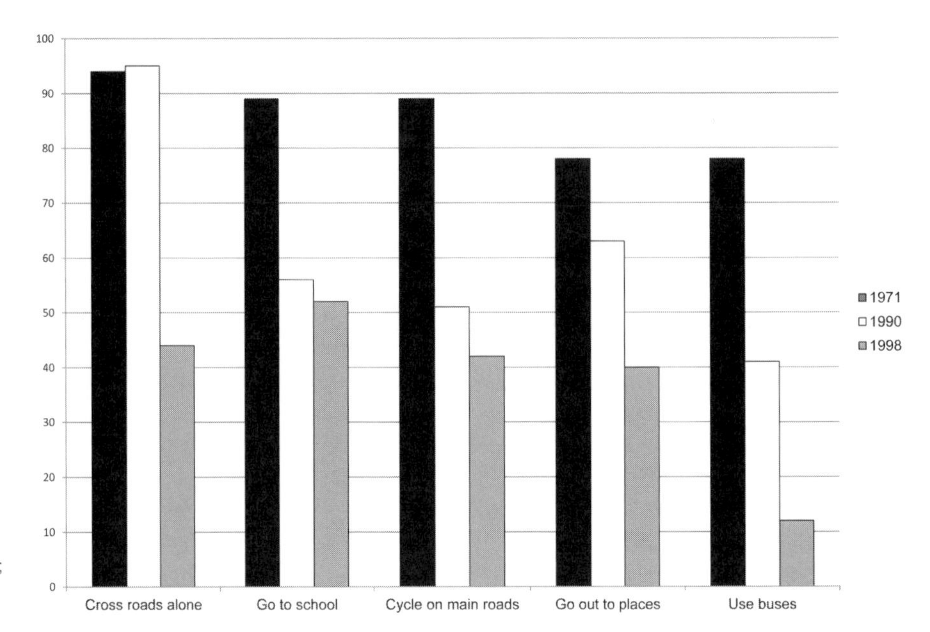

Figure 10.3 Proportion of British 10- and 11-year-olds able to undertake activities unaccompanied

Source: Hillman et al (1990); O'Brien et al (2000); Barker (2006, p50)

With increasing use of cars to access the spaces that children use, the average distance to school has increased over recent decades. In the UK, the average distance from home to primary school increased by 19 per cent from the early 1990s to 2005. Children are also now travelling much longer distances to visit friends. This is partly a response to the increasing size of school catchments, as local schools are amalgamated, or as parents choose the 'best' school. If a child travels 5km to school, and their best friend lives 5km on the

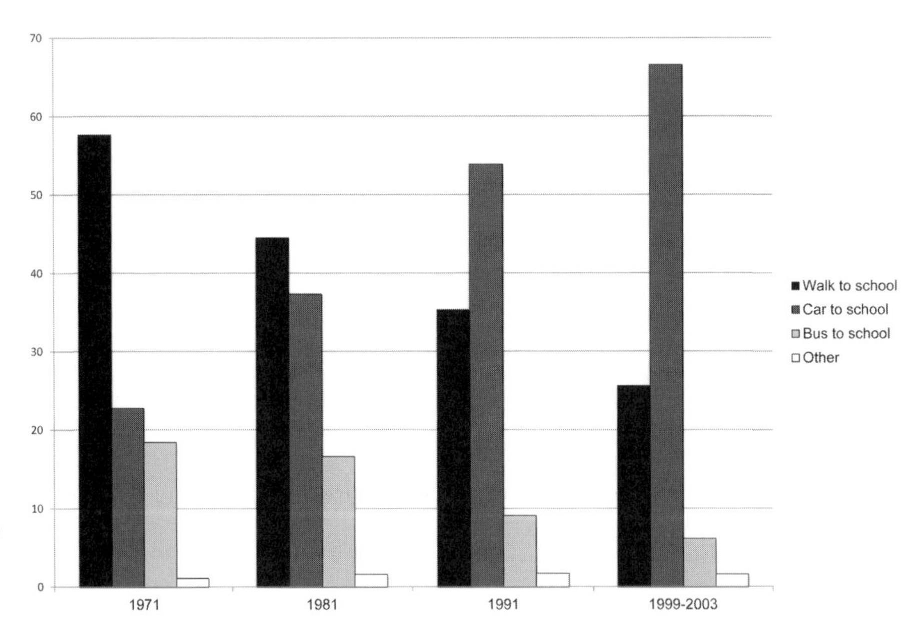

Figure 10.4 Mode of travel to school for children aged five to nine in Australia

Source: van der Ploeg et al (2008)

other side of the school, this makes it much more likely that car transport will be required to visit their friend. In contrast, if children attend a local school, and most live within 1km of the school, then their friends are more likely to be accessible by walking and cycling. Children's mobility is now not determined by their age or physical ability as much as it is by the ability of their parents to provide car transport for them. The following quote from a parent interviewed in a Norwegian study is apposite:

> Yes, we drive back and forth. When I was a kid, I was in the scout group, I did gymnastics and everything, but we had no transport. Then we had to cycle. But now we have another situation. Now I feel more time pressure. We have to be on the alert, so to speak ... Many families have several activities on the same day. (mother, aged 43, cited in Skar and Krogh, 2009, p346)

Public awareness of reductions in CIM has been heightened by newspaper coverage referring to 'the bubble-wrap generation' or to children's loss of their 'right to roam' (Cadzow, 2004; Derbyshire, 2007). In an article in the English newspaper the *Daily Mail*, the loss of children's freedoms was depicted using a map showing the play ranges of children in the city of Sheffield over four generations, starting with the great grandfather in 1919, who was allowed to walk 6 miles (9.7km) to go fishing. With subsequent generations, the range shrank, so that in 2007, the great grandchild is only allowed to walk 300 yards (274m), to the end of his street (Derbyshire, 2007).

Reasons for Variability in Children's Independent Mobility

There are complex and interrelated reasons for the differences in children's independent mobility between different nations, as well as the reduction in this mobility over recent decades. Children in many European countries tend to have higher levels of independent mobility due to differences in urban form and transport, local differences in the planning of housing estates, and differences in the culture or community ethos regarding children.

If we compare some of the European cities (e.g. German cities) or Japanese cities, where children have higher levels of independent mobility, with countries such as the US or Australia, we can identify some possible reasons for children's lower levels of freedom. Australian and US cities are more dispersed, with longer distances for children to travel to school, to shops and to a range of service spaces. Given the longer distances, parents are more likely to have to provide car-based access.

Child-friendly transport modes are walking, cycling and public transport

Child-friendly transport modes are walking, cycling and public transport. These are child friendly because children can use them independently, without an adult, and when adults use these modes, the city does not become less child friendly. The more that adults use these modes, the more likely children will feel safe using them because of the safety in numbers effect (Jacobsen, 2003). Public transport in many Japanese and German cities is frequent, safe and reliable. In contrast, in most Australian and many US cities, public transport is poorly regarded, and children seen travelling on public transport, especially in the evenings, would be regarded as 'out of place'. There is also far less provision for pedestrians and cyclists, as well as a culture in which public transport users, pedestrians and cyclists are seen as inferior to car occupants. In many cities in The Netherlands, Germany, Denmark and Japan, speed limits are restricted to levels that are safer for pedestrians and cyclists (and, hence, children as well). Speed limits are often 30km/hour or as low as 20 or 15km/hour. Some streets in Denmark are designated and clearly marked as areas where children play (Tranter and Doyle, 1996). In Berlin, over 72 per cent of the streets are traffic calmed with speed limits of 30km/hour or less (Pucher et al, 2009).

As well as these differences in urban form and transport, cultural differences may also explain the differences in CIM. In Japan and Germany 'a whole of community ethos has developed ... which actively looks out for children using public spaces' (Thomson, 2009). In Germany, as Hillman et al (1990) argue, there is a greater collective responsibility for other people's children. If a child needs assistance, or is misbehaving, German adults are likely to intervene, even if the child is not known to them. Adults in Australia, the UK or New Zealand are far less likely to get involved, as they fear that their motives might be misconstrued.

Tim Gill (2007), author of *No Fear: Growing Up in a Risk Averse Society*, tells a heart-rending story that indicates the impact of the lack of a culture that actively looks out for children, in this case in the UK:

> Abigail Rae was a toddler who wandered out of her nursery, and drowned in a nearby pond. A male passerby told her inquest he had spotted her in the street, and been concerned. But he hadn't saved her life by stepping in, he said, for fear of looking like he was abducting her. (Reported in *The Guardian*, Aitkenhead, 2007)

We can also ask why the levels of CIM have declined so dramatically over the last few decades, particularly in Canada, the US, the UK, Australia and New Zealand. Again there are several factors that in combination are likely to have influenced the decline. Some relate to changes in the features of the spaces that children occupy, while others relate to cultural changes.

The dominance of a neoliberal policy agenda has led to the winding back of public investment in cities, particularly in the UK, the US, Australia and New Zealand (Gleeson and Low, 2000). One manifestation of this is the policy of amalgamating and closing schools, with the stated objective of saving money for education departments or, alternatively, providing 'better education' in larger schools. Such a focus is based on a narrow view related to the efficiency of the formal educational provision (Steen, 2003), rather than recognizing that significant learning occurs outside the classroom, even on the journey to school (see Chapter 4). In many US cities, children have been forced to travel long distances and mothers average more than an hour per day driving children around (Beaumont and Pianca, 2000). Some policies in US cities are biased towards forcing local school boards to build new schools rather than renovating older ones, even if renovation is cheaper. The effect of such policies is to make it more difficult for many children to walk or cycle to school, and to increase the numbers of children being driven. This then compounds traffic congestion at the school gate, which puts more pressure on other parents to drive their children to school in order to protect their own children from traffic danger.

In many US cities mothers average more than an hour per day driving children around

When schools close, this places further pressure on local shops and services, as local shops often depend upon the business from teachers, parents and children attending nearby schools. When the local shops and post offices close, more traffic is generated as suburbs become dormitory areas. The extra traffic adds to parental concerns about safety and reduces the number of children allowed to walk around the local streets. This then increases fears of 'stranger danger': when everyone drives rather than walks or cycles, the streets become seen as lonely, deserted and dangerous places, where no one is around to provide passive surveillance for children, particularly adults who know their neighbours' children.

Passive surveillance is also reduced by the design of new suburban estates. Streetscapes are now more likely to be dominated by double or triple garage doors, so that there is limited window frontage onto the street. Garage doors are often operated with remote controls, so that residents do not even get out of their cars before they are inside their own houses. This also reduces the opportunity for social connection between neighbours. The garage frontages of houses are very different from the front porches or verandas that were typical of older houses. Front porches are now making a comeback in New Urbanist developments, such as Tullimbar, on the south coast of New South Wales. Front porches are not simply symbolic; they promote social interaction and surveillance (Hess, 2008) and encourage people to interact with the street, providing a stronger sense of community and connection that is important for a child-friendly neighbourhood.

As well as physical changes to the urban areas in which children live, changes to the community culture have influenced children's mobility. The most important of these relate to the changes in the pace and complexity of urban life, the increased competitive pressure to be 'good parents', the development of a risk-averse culture (see Chapter 4) and the concept of 'social traps'.

Time pressure is a reason why parents drive their children to school

Time pressure is often listed as a main reason why parents drive their children to school (Mackett, 2002). Many families today have highly constrained time schedules, with rigid work times and with demands on children's time from extra-curricular activities. A typical day for many parents starts with driving their children to school, followed by driving to work, picking up children from after-school care, driving home for dinner, then out to the shops, and then to soccer training, to the gym, to yoga and then home. Many parents feel that there is not enough time to complete all the activities.

Part of the reason for this increased time pressure comes from the parents competing with each other to give their children the best opportunities. This relates to globalization and increased competition and uncertainty in the workplace, which means that parents are more anxious about preparing their children for adult life (Honoré, 2009). Parents are also often caught in social traps, which force them to behave in ways where the collective impact of individual decisions is damaging to children – for example, by eroding their independent mobility.

The concept of 'social traps' or 'social dilemmas' (Hallsworth et al, 1995; Black, 1997; Kitamura et al, 1999; Tranter, 2006) can be applied to children's licences of independent mobility. Many parents feel 'trapped' into driving their children because the majority of other parents are contributing to a dangerous situation by driving their children to school. Parents may also fear social alienation if they do not conform to the social pressure to be a 'good parent'.

Many parents deem that cars are being used unreasonably for school trips, but cite the extra traffic around schools as a reason for driving their own child to school (Tranter and Pawson, 2001). School buses can alleviate this congestion (see Figure 10.5). Parents who let their children walk or cycle to school can feel social censure from other parents who believe that they are abrogating their responsibilities to protect their children from danger. The conceptualization of children in many modern Western societies is now dominated by a view of vulnerable children who must be protected from the dangers of modern society. The notion of children as capable social actors has been pushed aside in an era where children are driven everywhere to 'keep them safe'.

Not all actions that restrict children's freedom are based on a concern for their safety or protection. Some policies are designed more to reduce any disturbances caused by (unruly) children or youth. These include 'zero tolerance' policies for young people in public space or enclosed shopping centres (Malone, 2001). Whitzman and Pike (2007) also identify institutional barriers to CIM, including laws relating to 'no cycling, skateboarding, or ball

Figure 10.5 A school bus collecting children from John W. Ross Elementary School, Washington: Note how the bus parks diagonally across the road to prevent cars from overtaking

Source: Paul Tranter

playing signs in parks' and local government fines for chalk graffiti or tree climbing (Louv, 2005; Gill, 2007).

Another important factor that is often ignored in explaining the decline in CIM is the availability of cheap energy for transport. Parents can only afford to drive their children to school, to sport, to maths tutoring and to their friends' houses if the cost of providing that transport is affordable. For the last several decades parents in modern Western societies have had access to cheap energy in the form of a reliable supply of oil, combined with high employment. This has meant that car ownership is widely available for most households. Indeed, in many countries, particularly in suburban locations, it is the norm to now have two or more cars per household. Access to the second car is often justified by the argument that it is needed to drive the children to school. In the final chapter, we explore the likely opportunities that might come for children from the phenomenon of 'peak oil': the peaking and consequent decline in global oil production, which is likely to signal the end of cheap oil (Tranter and Sharpe, 2007, 2008).

Strategies to Increase Children's Independent Mobility

If we assume that CIM is of value, and that it is worth reclaiming the levels of CIM that children enjoyed only a few decades ago, we can identify ways of transforming our cities so that our children change from 'battery reared' to 'free range' (Whitzman and Pike, 2007; Skenazy, 2009). We outline below three broad strategies that are likely to improve children's freedom to playfully explore their cities. The first of these involves major changes to urban transport and the location of schools, shops and services. The second involves finer detail changes to urban neighbourhoods. The third involves cultural change: changes in social values about parenting, the pace of life and individualism. National or state government departments usually achieve the first of these; the second, local government, and the third can be effected at any level, from local communities or streets through to federal government policies.

To enable children to move freely around their neighbourhoods and cities we must invest in child-friendly transport modes (walking, cycling and public transport). Denmark, Germany and The Netherlands have shown that a concerted effort to enhance these modes can transform their use by urban residents (Pucher and Buehler, 2008). Copenhagen is an excellent example, where walking and cycling rates have increased dramatically over the last few decades, and car use in the central city areas has declined. This has been achieved by a process of progressively removing car access and car parking, and giving this back to people for walking, cycling, social interaction and playing. Just a few decades ago, Copenhagen was similar to many UK cities in terms of the dominance of the car in urban transport. Now over 30 per cent of commute trips are made by bicycle. Similar success has been achieved in South American cities such as Curitiba (Brazil) and Bogotá (Columbia), where public transport ridership has been dramatically increased through the use of low-cost innovative bus rapid transit systems. Box 10.2 provides a case study of a bold step towards a child-friendly city.

We must invest in child-friendly transport modes

As well as increasing the share for child-friendly transport modes, it is also important to ensure that schools, shops and services are located close to where children live. In research on children, one variable that consistently influences the likelihood of children going to and from school alone is distance from the school. For primary school children, the proportion walking to school starts to fall as distance increases beyond half a kilometre from school, and falls very rapidly beyond 1.5 km. In the UK, beyond 1 mile (1.6 km), the car becomes the dominant mode (Barker, 2006). Thus, if we can keep local schools open, rather than close or amalgamate them, this will enhance CIM.

At the level of local neighbourhoods, CIM can be enhanced through policies that limit the speed and the volume of traffic – for example, the creation of Home Zones in the UK (see Figure 10.6). Speed restrictions and traffic calming have been shown to be effective at reducing injury and pollution levels and encouraging more walking and cycling (Garrard, 2008). A common barrier to the introduction of speed reduction measures is the perceived increase in travel time and cost. However, such increases are usually extremely minor (9 seconds per kilometre travelled in one study), and such small travel time benefits also 'come at substantial cost in terms of the health and well-being of individuals' (Garrard, 2008, p10).

The sense of connection that children feel when walking or cycling around their suburb is also important. Some housing trends should be avoided if we want to make streets welcoming places for adults and children, and places where a sense of passive surveillance can be achieved. An example of what *not* to do in a residential street is presented in Figure 10.7. This shows a street in

Box 10.2 Three days to a more child-friendly city

The city of Curitiba in Brazil provides an inspiring story for those who want to make a city more child friendly. During the 1960s and 1970s, part of the vision to facilitate massive growth in car ownership involved destroying Curitiba's central city street, Rua Quinze, by building a new motorway overpass (Khayesi et al, 2010).

Curitiba's future was swung away from meeting the needs of the car when a new mayor – Jaime Lerner – was appointed in 1971. Lerner believed that the city should go in a completely different direction: the street should become a pedestrian mall. Knowing that business owners opposed his idea, Lerner ordered the pedestrianization be completed in 48 hours. It took only slightly longer, and in just three days the municipal authorities had installed pavers, lighting, planters and furniture. Storeowners were soon convinced of the wisdom of the move when their turnover showed a sharp increase. They even called for an extension of the traffic-free area.

Motorists, however were upset by the move, and planned a demonstration – to drive their cars into the new mall on a Saturday. Upon learning of this, Lerner 'contrived an unbreachable defense' (Lubow, 2007). He enlisted the support of hundreds of children, armed with boxes of paint and rolls of paper, and with encouragement from teachers, the children sat and drew pictures in the newly traffic-free street: It was to say: 'This is being done for children and their parents – don't even think of putting cars there' (Lubow, 2007). The pedestrianized shopping area has since expanded so that 15 blocks in the city are now car free.

Not only were children involved in the defence of this bold move to take the city back from the car, their independent mobility was enhanced through a policy that involved taking a risk with a radical step. Such steps are now seen as good sense in the city of Copenhagen, where the dominance of the car is being incrementally challenged by making space for people.

an outer-Melbourne suburb, where children would experience the feeling of being alienated rather than welcome. Several houses in the street have shutters and/or high front fences. While these are thought to provide security benefits for the householder, they provide little security for the local community.

Figure 10.6 Designating Home Zone areas is one approach to reclaiming streets for children, although removing cars is still a challenge

Source: Claire Freeman

Figure 10.7 A suburban street in Melbourne: The shutters and high spiked front fences not only cut off interaction between the street and its residents, but send strong messages of preferred non-contact

Source: Paul Tranter

The third broad approach to increasing CIM is cultural change, or changes in social values. Significant changes are required before we can expect to see changes in children's freedoms: a challenge to the dominance of speed and productivity in our lives; a switch to an appreciation of collective responsibility, rather than individualism; and a challenge to the notion of a 'good parent'.

Modern societies have developed a culture that emphasizes speed, efficiency and saving time: 'children are not born obsessed with speed and productivity – we make them that way' (Honoré, 2004, pp216–217). Parents admit that the most common phrase they use when bringing up their children is 'Come on! Hurry up!' Many adults also believe that children playing are wasting their time, so the tendency is to fill their time with stimulating activities that they can learn from. Yet, there is little evidence that this extra stimulation is effective. Danish children, who start formal schooling at the age of six or seven, perform better than British children who start two years earlier. Children learn best when they can learn at their own pace.

Children are not born obsessed with speed and productivity – we make them that way

Urban residents also need to become more aware of the collective impacts of individual decisions. These are evident when parents aim to keep their children safe by driving them to school in order to protect them from the traffic created by parents who drive their children to school. A collective agreement to look after everyone's children is needed to break out of these social traps. This can be achieved with safe routes to school programmes and plans, which might include strategies such as the walking school bus (WSB), an idea first proposed by an Australian innovator (Engwicht, 1992). Although this idea has now become more formalized than Engwicht first intended, it has provided a bridge between car-based mobility and CIM for some children, particularly those whose parents are concerned about traffic and stranger danger. The WSB overcomes these dangers through adult supervision and the safety in numbers principle (Kingham and Ussher, 2006; May et al, 2008).

Part of a switch to understanding the collective impact of individual decisions involves a questioning of the dominant model of a 'good parent'. Typically, a 'good parent' is one who drives their children to every activity (school, sports, friends, shops), preferably in a 'safe' car such as a large 4WD. A multitude of stimulating extra-curricular activities is provided for their children, and they are also protected from any dangers that might arise through unstructured play. Parents strive to give their own children a competitive advantage in a competitive world (Thomson, 2009).

An alternative model of a good parent would involve letting children enjoy their childhood. One influential book on children's well-being suggests that we should value 'childhoods for children, rather than for adults' (Stanley et al, 2005, p166). These authors 'challenge the current cramming of childhood with prescribed and supposedly enhancing activities, without enough time for the child to enjoy, reflect and grow in his or her own special and spontaneous ways' (Stanley et al, 2005, p161). Thus, rather than attempting to prepare children for a successful adulthood in a consumerist world, we should be letting them develop their own resilience, and we can do this by letting them playfully explore their environment, or by supporting active transport by walking and cycling with children (see Figure 10.8).

Figure 10.8 Active transport is also important for children: Even walking in the rain in a crowded city street can be an adventure for children

Source: Paul Tranter

Reversing the Trend

Efforts to increase children's freedoms provide benefits for all city residents

To achieve a reversal in the current trend in many cities towards lower levels of CIM will require broad changes, including transformations in the culture of transport and the conceptualization of children. In order to increase children's independent mobility we will need to challenge the dominance of the car as a mode of transport, as well as the cultural perception that the car is a superior mode of transport – a view already formed in young children when they learn from their parents' behaviour that the car should be used even for very short journeys. Fortunately, models for enhancing use of the non-car modes already exist in many cities, Copenhagen being a well-known example. It is imperative that children are conceptualized as citizens in the 'here and now', rather than 'adults in training', and that children's views are included in policies to address declines in CIM. When given the chance, children usually prefer independent modes of travel. Given the value of CIM for children, parents and the wider environment and community, efforts to increase children's freedoms will provide benefits for all city residents, both now and in the future. It will also help to make our cities more resilient in the face of challenges such as global climate change, peak oil and economic crises. The following chapter turns to the issue of design as a strategy in making our cities better places for children.

References

Aitkenhead, D. (2007) 'Why we need to set our kids free', *The Guardian*, 3 November 2007, www.guardian.co.uk/lifeandstyle/2007/nov/03/familyandrelationships.family5

Barker, J. (2006) '"Are we there yet?": Exploring aspects of automobility in children's lives', PhD thesis, Department of Geography and Earth Sciences, Brunel University, London

Beaumont, C. and Pianca, E. (2000) 'Historic neighborhood schools in the age of sprawl: Why Johnny can't walk to school', National Trust for Historic Preservation, Washington, DC, available at www.preservationnation.org/issues/historic-schools/additional-resources/schools_why_johnny_1.pdf, accessed November 2010

Black, C. S. (1997) 'Behavioural dimensions of the transport sustainability problem', PhD thesis, University of Portsmouth, UK

Burdette, H. L. and Whitaker, R. C. (2005) 'Resurrecting free play in young children: Looking beyond fitness and fatness to attention, affiliation, and affect', *Archives of Pediatrics & Adolescent Medicine*, vol 159, no 1, pp46–50

Cadzow, J. (2004) 'The bubble-wrap generation', *Sydney Morning Herald Good Weekend Magazine*, 17 January, p18–21

Chawla, L. (2002) *Growing Up In an Urbanizing World*, Earthscan/UNESCO, London

Collins, D. and Kearns, R. (2001) 'The safe journeys of an enterprising school: Negotiating landscapes of opportunity and risk', *Health and Place*, vol 7, no 4, pp293–306

Crider, L. B. (2003) '"Safeways to schools" for Florida's school children', Paper presented to the Walk21 Conference, Portland, OR

Derbyshire, D. (2007) 'How children lost the right to roam in four generations', *Daily Mail*, 15 June 2007, www.dailymail.co.uk/news/article-462091/How-children-lost-right-roam-generations.html

Engwicht, D. (1992) *Towards an Eco-City: Calming the Traffic*, Envirobook, Sydney

Fjortoft, I. and Sageie, J. (2000) 'The natural environment as a playground for children: Landscape description and analyses of a natural landscape', *Landscape and Urban Planning*, vol 48, no 1–2, pp83–97
Fotel, T. and Thomsen, U. (2004) 'The surveillance of children's mobility', *Surveillance and Society*, vol 1, no 4, pp535–554
Garrard, J. (2008) *Safe Speed: Promoting Safe Walking and Cycling by Reducing Traffic Speed*, The Heart Foundation, City of Port Phillip and City of Yarra, Melbourne
Gilbert, R. and O'Brien, C. (2005) *Child- and Youth-Friendly Land-Use and Transport Planning Guidelines*, Centre for Sustainable Transportation, Toronto, Canada
Gill, T. (2007) *No Fear: Growing Up in a Risk Averse Society*, Calouste Gulbenkian Foundation, London
Gleeson, B. and Low, N. (2000) *Australian Urban Planning: New Challenges, New Agendas*, Allen and Unwin, Sydney
Hallsworth, A. G., Black, C. S. and Tolley, R. (1995) 'Psycho-social dimensions of public quiescence towards risks from traffic generated atmospheric pollution', *Journal of Transport Geography*, vol 3, no 1, pp39–51
Hart, R. (1979) *Children's Experience of Place*, Irvington, New York, NY
Hess, P. (2008) 'Fronts and backs: The use of streets, yards, and alleys in Toronto-area New Urbanist neighbourhoods', *Journal of Planning Education and Research*, vol 28, no 2, pp196–212
Hillman, M., Adams, J. and Whitelegg, J. (1990) *One False Move: A Study of Children's Independent Mobility*, Policy Studies Institute, London
Holbrook, A. (2009) *The Green We Need: An Investigation of the Benefits of Green Life and Green Spaces for Urban Dwellers' Physical, Mental and Social Health*, Nursery and Garden Industry Australia, Nursery and Garden Industry Australia and SORTI, University of Newcastle, Sydney
Honoré, C. (2004) *In Praise of Slow: How a Worldwide Movement Is Challenging the Cult of Speed*, Orion, London
Honoré, C. (2009) *Under Pressure: Putting the Child Back in Childhood*, Vintage Canada, Toronto
Horelli, L. (2001) 'A comparison of children's autonomous mobility and environmental participation in northern and southern Europe: The cases of Finland and Italy', *Journal of Community & Applied Social Psychology*, vol 11, no 6, pp451–455
International Center for Technology Assessment, (2000) *In-Car Air Pollution*, CTA, www.icta.org/doc/In-car%20pollution%20report.pdf, accessed 15 September 2004
Jacobsen, P. (2003) 'Safety in numbers: More walkers and bicyclists, safer walking and bicycling', *Injury Prevention*, vol 9, no 3, pp205–209
Kegerreis, S. (1993) 'Independent mobility and children's mental and emotional development', in M. Hillman (ed) *Children, Transport and the Quality of Life*, PSI Press, London, pp28–34
Khayesi, M., Monheim, H. and Nebe, J. (2010) 'Negotiating "streets for all" in urban transport planning: The case for pedestrians, cyclists and street vendors in Nairobi, Kenya', *Antipode*, vol 42, no 1, pp103–126
Kingham, S. and Ussher, S. (2006) 'An assessment of the benefits of the walking school bus in Christchurch, New Zealand', *Transportation Research, Part A: Policy and Practice*, vol 41, no 6, pp502–510
Kitamura, R., Nakayama, S. and Yamamoto, T. (1999) 'Self-reinforcing motorization: Can travel demand management take us out of the social trap?', *Transport Policy*, vol 6, pp135–145
Kytta, M. (2004) 'The extent of children's independent mobility and the number of actualized affordances as criteria for child-friendly environments', *Journal of Environmental Psychology*, vol 24, no 2 pp179–198

Louv, R. (2005) *Last Child in the Woods: Saving Our Children from Nature-Deficit Disorder*, Algonquin, Chapel Hill

Lubow, R. (2007) 'The Road to Curitiba', *The New York Times*, 20 May 2007

Mackett, R. L. (2002) 'Increasing car dependency of children: Should we be worried?', *Municipal Engineer*, vol 151, no 1, pp29–38

Malone, K. (2001) 'Editorial: Children, youth and sustainable cities', *Local Environment*, vol 6, no 1, pp5–12

May, M., Tranter, P. and Warn, J. (2008) 'Towards a holistic framework for road safety in Australia', *Journal of Transport Geography*, vol 16, no 6, pp395–405

Moore, R. C. (1986) *Childhood's Domain: Play and Place in Child Development*, Croom Helm, London

Morris, J., Wang, F. and Lilja, L. (2001) 'School children's travel patterns – a look back and a way forward', in *Papers of 24th Australasian Transport Research Forum: Zero Road Toll: A Dream or a Realistic Vision*, Tasmanian Department of Infrastructure Energy and Resources, Hobart

O'Brien, C. (2003) 'Transportation that's actually good for the soul', *National Center for Bicycling and Walking (NCBW) Forum (Canada)*, vol 54, pp1–13

O'Brien, M., Jones, D., Sloan, D. and Rustin, M. (2000) 'Children's independent spatial mobility in the urban public realm', *Childhood*, vol 7, no 3, p257

Offer, S. and Schneider, B. (2007) 'Children's role in generating social capital', *Social Forces*, vol 85, no 3, pp1125–1142

Pucher, J. and Buehler, R. (2008) 'Making cycling irresistible: Lessons from The Netherlands, Denmark and Germany', *Transport Reviews*, vol 28, no 4, pp495–528

Pucher, J., Dill, J. and Handy, S. (2009) 'Infrastructure, programs, and policies to increase bicycling: An international review', *Preventive Medicine*, vol 50, no S1, ppS106–S125

Rissotto, A. and Tonucci, F. (2002) 'Freedom of movement and environmental knowledge in elementary school children', *Journal of Environmental Psychology*, vol 22, no 1–2, pp65–77

Sandqvist, K. (2002) 'How does a family car matter? Leisure, travel and attitudes of adolescents in inner city Stockholm', *World Transport Policy and Practice*, vol 8, no 1, pp11–18

Shaz, K. and Corpuz, G. (2008) 'Serving passengers – are you being served?', Paper presented to the 4th Annual PATREC Research Forum, Edith Cowan University, Perth, WA

Skar, M. and Krogh, E. (2009) 'Changes in children's nature-based experiences near home: From spontaneous play to adult-controlled, planned and organised activities', *Children's Geographies*, vol 7, no 3, pp339–354

Skenazy, L. (2009) *Free-Range Kids: Giving Our Children the Freedom We Had Without Going Nuts with Worry*, John Wiley & Sons Inc, San Francisco, CA

Stanley, F., Richardson, S. and Prior, M. (2005) *Children of the Lucky Country? How Australian Society Has Turned Its Back on Children and Why Children Matter*, Macmillan, Sydney

Steen, S. (2003) 'Bastions of mechanism, castles built on sand: A critique of schooling from an ecological perspective', *Canadian Journal of Environmental Education*, vol 8, Spring, pp191–203

Thomson, L. (2009) *'How Times Have Changed': Active Transport Literature Review*, VicHealth, Melbourne

Timperio, A., Crawford, D., Telford, A. and Salmon, J. (2004) 'Perceptions about the local neighborhood and walking and cycling among children', *Preventive Medicine*, vol 38, no 1, pp39–47

Tranter, P. J. (1996). 'Children's independent mobility and urban form in Australasian, English and German cities', in D. Hensher, J. King and T. Oum (eds) *World Transport*

Research: Proceedings of the Seventh World Conference on Transport Research, vol 3: Transport Policy, World Conference on Transport Research, Sydney, pp31–44

Tranter, P. J. (2006) 'Overcoming social traps: A key to creating child friendly cities', in B. Gleeson and N. Sipe (eds) *Creating Child Friendly Cities: Reinstating Kids in the City*, Routledge, New York, NY, pp121–135

Tranter, P. (2010) 'Speed kills: The complex links between transport, lack of time and urban health', *Journal of Urban Health*, vol 87, no 2, pp155–166

Tranter, P. J. and Doyle, J. (1996) 'Reclaiming the residential street as playspace', *International Play Journal*, vol 4, pp81–97

Tranter, P. J. and Pawson, E. (2001) 'Children's access to local environments: A case-study of Christchurch, New Zealand', *Local Environment*, vol 6, no 1, pp27–48

Tranter, P. J. and Sharpe, S. (2007) 'Children and peak oil: An opportunity in crisis', *International Journal of Children's Rights*, vol 15, no 1, pp181–197

Tranter, P. and Sharpe, S. (2008) 'Escaping Monstropolis: Child-friendly cities, peak oil and Monsters, Inc.', *Children's Geographies*, vol 6, no 3, pp295–308

van der Ploeg, H., Merom, D., Corpuz, G. and Bauman, A. (2008) 'Trends in Australian children traveling to school 1971–2003: Burning petrol or carbohydrates?', *Preventive Medicine*, vol 46, no 1, pp60–62

Weller, S. and Bruegel, I. (2009) 'Children's "place" in the development of neighbourhood social capital', *Urban Studies*, vol 46, no 3, pp629–643

Whitzman, C. and Pike, L. (2007) *From Battery-Reared to Free Range Children: Institutional Barriers and Enablers to Children's Independent Mobility in Victoria, Australia*, Report for Australasian Centre for Governance and Management of Urban Transportation, Australasian Centre for Governance and Management of Urban Transportation (GAMUT), Melbourne

11
Design

> We must learn how to know children – and the elderly – in the building of our cities and streets. For only then will we honor the child we once were – and the elder we are yet to become. It is then we will discover the creative potency of continuing the past with the future ... and with wisdom. (Engwicht, 1999, p6)

Design Matters

In 2008 there was a significant event in the history of global society: for the first time more people lived in urban (3.3 billion) than rural areas. By 2030, the number may be 5 billion (United Nations Population Fund, 2007). Although most of this growth is occurring in developing countries, urban growth, nonetheless, presents challenges for developed countries. These challenges include the need to provide for growth, to improve the existing urban fabric and to address decline, all three of which can occur simultaneously in different parts of the urban fabric. In this chapter we look at the design of cities and city form in relation to children's needs. New urban forms and development principles such as New Urbanism, liveable neighbourhoods, transport-oriented development and green cities offer the potential for creating much better urban environments. However, the unfortunate reality is that most children live in parts of the city that do not benefit from more enlightened thinking about urban form. Although some cities and parts of cities do offer good environments for children, many are blighted by pollution, heavy traffic, poor-quality buildings, alienation, decay, sprawl and isolation. The problems are not simple and neither are they purely economic. Brendan Gleeson (2006), writing on what he calls 'toxic cities', explains that while the Australian cities he writes about are economically wealthier, its children are 'fatter, sicker and sadder', a description that can also be applied to children in cities beyond Australia. The prevalence and ongoing growth of new poorly planned suburbs, together with the urgent need to retrofit the city to address existing problems in often deteriorating city environments, presents major challenges to urban planners and designers. The fact that new growth is perpetuating and continuing to create additional strata of unsuitable current and future environments for children is of particular concern.

Physical form influences children's ability to socialize

The physical form of the places in which children live and the buildings that they use matter. Physical form influences children's ability to socialize, their health, access to services, the well-being of their community and neighbourhood, their access to play spaces, independence, safety and their ability to enjoy their childhood. The actions of planners, designers, architects, engineers, landscape architects and other land management professionals are central to the process of urban design; yet few of these professions pay attention to the needs of children in urban design. Urban design is a rapidly growing sphere of influence and has been influential in recent developments towards making cities more liveable:

> Urban design: Is about making connections between people and places, between public and private space, between the natural and built environment, between movement and urban form, and between the social and economic purposes for which urban space is used. (Ministry for the Environment, 2002, p5)

Is there a way to design cities so that they support children's well-being?

Although urban designers may not consider children as an intentional component in design, fortunately, good urban design can benefit children. In this chapter we look at the physical challenges facing cities, cascading from the city scale as it relates to the overall shape of the city to the neighbourhood scale and down to the level of individual buildings. The subject of how cities can be improved for children at all these levels is addressed. New urban forms and initiatives are examined with reference to their appropriateness for children. Improving the urban environment for children comprises two parallel strands: physical improvement through better urban design, and the inclusion of children in the process of urban planning (discussed in Chapter 12). This chapter addresses the following question: is there a way to design cities so that they support children's well-being? With this question in mind the chapter starts by looking at city form and urban design principles, particularly how these can apply in cities experiencing major challenges arising from both growth and decline.

Good Urban Design

Cities vary significantly in population from a few thousand to many millions, and in size from small compact towns to the sprawling megalopolises of the US's northern seaboard, Germany's Rhine-Ruhr and Tokyo's metropolitan areas. Yet, despite these huge disparities, what makes for a successful city in design terms transcends these size differences. Curitiba in Brazil – with its population of nearly 2 million and location in a developing country – seems an unlikely candidate for worldwide planning acclaim; yet its transport system, used by 85 per cent of its inhabitants, is drawn on internationally to illustrate excellence in sustainable transport planning. Its planning also acts as a showcase of ecological and humane urbanism (Local Governments for Sustainability, 2010) especially commendable given that Brazil is not one of the world's more wealthy countries, ranking 65th in terms of per capita gross domestic product (GDP) (World Bank, 2005). Awarded the Globe Sustainable City award in 2010, Curitiba shows that excellence in planning design, combined with city managers committed to better cities, can produce highly effective results. In Europe, the Swedish city of Malmö was also nominated for the 2010 award for its transformed and rejuvenated heavy industrial core, indicating that positive urban design can also be achieved in older established cities experiencing decline. One of the aims of the urban regeneration in Malmö was to provide 'a pleasant living environment for everyone in Malmö. Those who live in

Malmö will have access to safe and secure homes. All children will be able to play outside in a healthy and inspiring environment' (Malmö City, Environment Department, 2009, p8).

The Globe awards are given for good urban sustainable development principles, which are decided by adults. Less well recognized are the urban design principles that children value. In 1994 the World Wide Fund for Nature's Let's Win Back Our Cities congress was held in Bologna. Over 300 children attended from around 100 Italian schools (over 1000 schools were involved in related projects) at which they devised the following *Children's Manifesto*. The manifesto identifies what children consider to be the key elements in city design.

The Bologna *Children's Manifesto* (see Box 11.1) is significant as it is one directly created by children themselves. More usually, guidelines on city design

Box 11.1 *Children's Manifesto,* Bologna, 1994

We need:

- gathering places to be able to interact with nature, roam freely, trust, have different experiences, decide how to spend free time.

In our cities we want:

- children's spaces, spaces to play close to home, streets where cars go slowly, community-managed open space, space that can be changed, sports facilities, children's theatres, meeting places, places to communicate, to meet different groups, to be alone and to be with others.

We want green spaces with:

- natural elements, trees, bush, tall grass, fruit trees, no play structures, lawns, the ability to build huts and hide, spaces that are easy to reach, used by lots of people, safe from traffic, lots of water, paths and slopes, available in every neighbourhood.

In our schools we want:

- courtyards to play and meet friends, to plant gardens, buildings easy to get to, a beautiful, colourful school in the middle of a garden.

To save us from suffocating traffic we want:

- to be able to move safely at all hours, streets with bike lanes, quiet no-traffic zones, no cars on footpaths, bike parks, ramps, accessible public transport, smaller buses, green barriers, friendly, understandable street signs.

To decide our future we want:

- children's councils which make decisions, regular meetings with administrators, to be informed, to be listened to; we want to be able to contribute to decisions that affect us.

Source: adapted from Francis and Lorenzo (2006, pp229–230)

are created 'for' children, though children may be involved in their creation. Table 11.1 identifies design principles that focus specifically on meeting children's needs, whereas Table 11.2 identifies general principles of good design developed by design organizations and individual city councils. What is most evident in a comparison of the two tables is the degree of commonality across the different sets of principles. Common principles clearly emerge – namely, participation, connectedness (physical and social), access to services, access through walkability, bikeability and public transport, green spaces, a sense of identity, safety and good-quality structures. The principles that prioritize children's needs accord with general good design principles: what works for children works for the wider public. What is interesting, though, is what is missing from these principles: the issues that currently seem to dominate urban planning. Nowhere do they mention providing for more rapid traffic through-flow and car parks, expanding suburban housing on the urban periphery, creating new shopping malls and big box shopping outlets, or agglomerating local services into larger more distant units. Urban planners and designers recognize good planning and design, yet seem to find it extraordinarily difficult to implement given some of the ongoing practices where the emphasis is frequently on immediate, reactive, *ad hoc*, least cost, short-term planning. However, in fairness, the challenges can be huge. The example of Perth, Australia, is used to illustrate some of the stark realities and more extreme pressures facing urban planning.

Table 11.1 Child-friendly urban design principles

UNICEF Child-Friendly Cities (UNICEF, 2009; Malone, 2006)	**New South Wales Commission for Children and Young People (2009)**	**Delft Manifesto (Childstreet, 2005)**	**Francis and Lorenzo (2006)**
Influence decisions about their city Express their opinion on the city that they want Participate in family, community and social life Receive basic services such as healthcare and education Drink safe water and have access to proper sanitation Be protected from exploitation, violence and abuse Walk safely in the streets on their own Meet friends and play Have green spaces for plants and animals Live in an unpolluted environment Participate in cultural and social events Be an equal citizen with access to every service	Agency: • independent access to community services; • engage actively in community. Safety and security: • make community public places safer; • increase ability of children to feel secure and connected to the community. Positive sense of self: • spaces that are fun, welcoming and supportive; • increase access to green space/natural areas; • influence decisions about their city.	Protection: • social safety and traffic safety. Walkability: • safe crossing to the other side of the street and space for walking. Cycleability: • safe crossing to the other side of the street and facilities for cycling. Criss-Crossability: • suitability to use the full width of the street. Enjoyability: • attractiveness and variation. Playability: • suitability for various activities.	Accessibility Mixed use and mixed users Sociability Small, feasible, flexible Natural, environmentally healthy Urban place and identity Places and opportunities for participation

Table 11.2 Urban design principles used by organizations and cities

Organization design principles			City design principles		
New Urbanism (undated)	**European Union Expert Group (Lloyd-Jones, 2004)**	**Resilient City. org (2010)**	**Melbourne City Council (2006)**	**Birmingham City Council (2001)**	**City of Los Angeles (2009)**
Walkability Connectivity Mixed-use and diversity Mixed housing Quality architecture and urban design Traditional neighbourhood structure Increased density Green transportation Sustainability Quality of life	High-quality places Vibrant, equitable economy Treats land as a precious resource Locates new developments with reference to the natural environment Mixed land use Sufficient density Green structure Quality public infrastructure and public transport Resource-saving technology Respects cultural heritage	Density, diversity and mix Pedestrians first Transit supportive Place-making Engaged communities Integrated natural systems Local sources Environmentally appropriate urban forms and infrastructure	Urban design principles Clear structure Connectedness Diversity Animation Authenticity Equity Good fit with people's intentions	Creating diversity Ability to move around easily Safe and private places Building for the future Emphasize local character	Inviting and accessible transit areas Reinforce walkability, bikeability and well-being Nurture neighbourhood character Bridge the past and the future Produce great green streets Generate public open space Stimulate sustainability and innovation Improve equity and opportunity for all Emphasize early implementation and simple processes Maintain long-term solutions Ensure connections

Note: Tables 11.1 and 11.2 have adapted the sets of principles, selecting those that are the most relevant. For the full list, the original sources should be consulted.

A Design Challenge: Can the City of Perth Be Child Friendly?

Much of the information in this discussion is derived from a stunning book entitled *Boomtown 2050* by Richard Weller (2009). Weller dedicates his book to 'my parents who bought their first home in the suburbs of Sydney in 1962 and raised happy kids'. Could happy kids be raised in suburban Perth today? Perth is now a city of 1.5 million people, whose size is predicted to double by 2050, requiring some 700,000 homes to be built. The city epitomizes sprawl as it extends over 120km in length. Its houses are getting bigger (the average size for new housing is 300 square metres) and its plot sizes smaller and, overall, more expensive, thus pricing houses outside the affordability range of most families. Typical home bank loans are five times annual incomes and house prices eight to nine times annual income. Cars rule in Perth. Each household has an average of 2.1 cars; each person makes an average of 3.1 car trips per day, with only 6.3 per cent of trips using walking or cycling compared to 35 to 45 per cent in Europe. It is estimated that 60 per cent of primary students live less than a 20-minute walk from school; yet most are driven on a daily basis.

Homes are built for families of a certain type, age and income; these are mainly larger homes, although 24 per cent of these family-size homes are occupied by individuals. The Perth sub-region is classed as one of the world's biodiversity hotspots; yet about 15,000ha of native vegetation was approved for clearance in 2005 to 2006. Its ecological footprint is seven times the world average. Ecologically, the city faces enormous challenges, isolated from the rest of Australia and situated in a region with meagre water resources. Its ecological precariousness led Tim Flannery, the influential Australian environmentalist, to predict that Perth will become the 21st century's 'first ghost metropolis'. In a nutshell, as a city, Perth fails to meet the good urban design principles identified in Tables 11.1 or 11.2.

What does all of this mean for children living in Perth? Many families will not be able to afford family housing. Where they are able to purchase family homes, they can find themselves working long hours to pay for the home and the cars needed for the commute, with both parents needing to work to pay for them. The average Perth worker works 41 hours per week; but to this has to be added the average 2-hour daily commute for workers from outer suburbs, which are mainly family housing areas. Up to 90 per cent of trips in the newer suburbs are by car; these suburbs lack local employment opportunities and are associated with lower service provision. The reliance on cars for transport and poor service distribution exacerbates children's inability to access both local services and the central city. During summer, temperatures can be very high: the mean maximum for January 2010 was 33.4°C, with a maximum of 42.9°C recorded (see http://weatherzone.com.au). These weather-related factors can make walking and cycling more difficult, especially given the low density and widely dispersed nature of urban facilities.

Does Perth have the capacity to create a good urban future for children?

Does Perth have the capacity to create a good urban future for children? Perth has a climate in which much of life can be lived outdoors. It has great beaches and ample green space, including some remnant native bushland and wetlands. Indoor space in homes is getting bigger, and outdoor space, as in gardens, is getting smaller. Land is available close to the city for higher-density housing, and as the Coin Street example (see Chapter 6) shows, well-designed city-centre housing can work for children. In Perth, higher-density housing is needed; but it would also have to be in areas with high amenity values, schools, health services, green space, sports facilities, etc. Higher-density housing in beach suburbs would enhance access to this excellent resource for more children; but these are areas where existing, invariably wealthier, residents would be likely to vociferously oppose higher densities. Perth's newer 'family' suburbs, located on the outer edge of the city, are built for cars and poorly served by public transport, exacerbating inaccessibility for children. These suburbs also tend towards monocultural populations and are lacking in both social and economic diversity. Although the suburbs may be well laid out with regard to plentiful green space, it is hard to imagine them as peopled, diverse and with an urban 'buzz'.

In order to provide a better suburban future, the Western Australian Planning Department launched its Liveable Neighbourhoods Initiative in 1997. This initiative promotes walkable neighbourhood catchments-pedsheds, mixed uses, a service core, interconnecting streets and safe routes, a diversity of housing, local identity and environmental sensitivity. However, realistic implementation and uptake by residents of reduced car use and more active local engagement seems to be a step too far for a city whose residents are wedded to a low-density car-oriented lifestyle. The ultimate concern for planners of a city such as Perth should be that its children are growing up with no other experience than that offered by their car-dependent low-density suburbs, and will see these as the norm in the future. Notwithstanding recent investments in high-quality public transport, children growing up in Perth are severely compromised by its low-density, dispersed, resource-hungry design. Fortunately, there are other more positive scenarios which anticipate better urban living.

City-Level Design

At the city level a number of design options for improved city form are available. Foremost among these are the green city and sustainable city forms, including the latest development, 'resilient cities' – that is, cities resilient in the face of urban crises, most notably the challenges of climate change and peak oil (Newman et al, 2009). These offer an alternative to the economically motivated and developer-driven development pattern that has led to the problems being experienced by cities such as Perth. Green and sustainable cities prioritize environments that respect nature through limiting environmental impact and resource use, and emphasize the need for residents to have contact with nature. Such cities are not only greener but more socially viable and supportive, encouraging social connectivity. Social contact cannot easily occur where homes are surrounded by high walls and entered courtesy of a remote control garage opener. Sprawl and low-density suburbs affect children's ability to connect; but so too can poorly designed high-density high rise. Sprawl cities such as Perth offer opportunities to fill in the gaps, creating higher densities within the existing urban fabric. Declining industrial cities offer opportunities to create amenity space on land vacated by industrial and commercial uses, benefiting children previously unable to access local green space (Ward, 1989). Queen Elizabeth Park in Vancouver, located near the city centre and currently Vancouver's second most visited park, includes extensive gardens and arboreta, yet began life as a basalt quarry. Retrofitting the urban environment to create better industrial, residential and mixed-use areas that incorporate green buildings, ecologically aware planning and green-space development is being taken seriously in parts of Europe. Cities are increasingly examining their sustainable, green and healthy credentials and seeking to enhance these. Some cities, such as Amsterdam, have a head start with their compact form, higher-density housing, tight urban boundary and strong accessibility. For others, the challenges are greater. While the overall city form is important as it sets the character of the city, for children the city is experienced at a smaller scale (the neighbourhood and street level), and it is here that some of the most exciting design initiatives are to be found.

At the neighbourhood level some exciting design initiatives are found

Neighbourhood Design

Foremost among contemporary initiatives for neighbourhood development is that of New Urbanism. While New Urbanism has its critics, it represents a validation of long-recognized principles of good development. These include mixed uses, varied housing types, attractive public realm, pedestrian-friendly streetscapes, defined centres and edges, public transport options, and services such as health centres, libraries and schools being located centrally at public transport intersections (United Nations Human Settlement Programme, 2009, p70). Buildings are at a scale that people can relate to. For children, scale is especially important: they interact with environments at a much smaller scale than do adults. Children notice things that are overlooked by adults and they are also affected in ways that adults aren't. For example, young children cannot see over parked cars or over garden walls, and in buggies/strollers they cross pavements at the level of car exhaust fumes. They find it hard to comprehend cities that are designed for use at car speeds rather than at walking speeds. Distance is an important part of scale; it governs – depending upon age and level of independence – children's access to shops, sport fields, friends and swimming pools. Scale and distance influence how far they can walk and how many roads have to be crossed to get to places. Scale influences whether children can interpret what is going on; high walls preclude them from being able to see whether anyone is at home or if the dog barking on the other side of the wall is fierce or not. Large roads are more difficult to judge in terms of whether they can be crossed safely than are smaller roads. It is harder to access a shopping mall to buy bread and milk than the corner shop. The more inhuman the scale, the harder it is for children to know people in their neighbourhood: the post/mail person, the shopkeeper, the local school teacher, the dog walker, the family at the end of the street and the old lady who sits on the corner. At the neighbourhood level, children need to relate socially and environmentally with people and places. A human scale of development allows this to happen as part of their daily lives. Good cities naturally subdivide into neighbourhood units that residents can relate to and which are places of supportive encounter.

The more inhuman the scale, the harder it is for children to know people

Higher-density housing developments are characteristic of developments in European countries such as The Netherlands and the UK, where the need for new housing, coupled with a shortage of land, has led to innovative high-density housing. Several such developments can be seen in Amsterdam. One of the best known is the redevelopment of the Eastern Docklands to create 8000 homes. Java Eiland, the first neighbourhood to be developed, comprises high-density housing around a pedestrianized core with internal courtyards and amenity spaces. The presence of bicycles with child carriers, basketball hoops, communal sandpits and its car-free core and walkways indicates the presence and consideration of children in planning this new development (see Figure 11.1). Proximity to public transport, water taxis and trams gives good access to wider city services. Quality and diversity is evident in the architecturally designed buildings, which further embrace Amsterdam's heritage through its building style and internal canal network. Java Eiland ticks most of the boxes for good urban design and, simultaneously, for good child-friendly design.

Figure 11.1 This development prioritizes pedestrian child-friendly spaces in its design (Java Eiland, Amsterdam)

Source: Claire Freeman

A similarly impressive Dutch development was featured in a UK TV series entitled *The Perfect Home* by architect and writer Alain de Botton (2006). In the series, together with Aaron Betsky, head of The Netherlands Architectural Institute, de Botton visits Ypenburg's new subsidized social housing development. Betsky describes the housing form as corresponding to 'human forms based on what a child would draw', giving it a 'human feeling'. He thus intimates that children intuitively know what works. When asked why the development seems to be successful again, he refers to children: 'All I know is whenever I come here, whether rain or shine, I always see kids playing and I often see people talking to one another.' In writing about the UK, Tim Gill (2007, p7) similarly maintains that 'the presence of children playing in the street can be seen as the litmus test of the level of community cohesiveness in a neighbourhood'. Good urban design allows children to be visible in the urban fabric.

A very different type of development is that of 'socially insulated neighbourhoods', designed to remove residents, including children, out of the public domain, the most commonly recognized example being gated developments. These are particularly controversial, especially with regard to children. There have been attempts to create both 'child-free' housing developments and housing developments where children can play and live safely away from the dangers of traffic and other perceived urban dangers. Although usually referred to as gated communities, the word 'community' is highly contested as they are customarily mono-social, mono-economic and closely regulated to ensure that they are only open to 'desirable' residents. The rapid growth of gated communities, especially in countries such as Brazil, the US, Australia and China, is an increasingly important urban form when looking at children's urban experience. While still relatively rare in Europe, in the US they are becoming the dominant form of new build. During the last 25 years (up to 2005), some 250,000 housing developments containing about 20 million housing units and 50 million people had been built in the US with an increasingly upward construction trend expected (McKenzie, 2005). The impacts of gated developments on children will be considerable, but are as yet largely unstudied.

Living Streets

Good street design is central to socially well-functioning neighbourhoods. The prime factors influencing sociability at the street level are scale and safety: scale as in the size of buildings and their relationship to the street, and safety as in pedestrian safety. In 2008, Hart looked at the relationship between traffic volumes on three UK streets and social relationships, replicating a similar San Francisco street study (Appleyard and Lintell, 1972). The UK study, like its San Francisco predecessor, found that the more traffic there was on streets, the fewer friends and social connections people had. Residents on heavy-traffic streets tend to have 75 per cent fewer friends on the street than those living on similar streets with less traffic (Hart, 2008). Families on 'heavy street' limit their children's time outdoors and report fewer social connections. Traffic, particularly heavy traffic, transforms streets from functioning as places of social encounter that bring people together into vehicular thoroughfares.

In an attempt to counter the transformation of streets into vehicular passageways, Engwicht (1999), on the opening page of his book *Street Reclaiming*, asks the reader to 'Imagine your street with 50 per cent less traffic. Imagine cars as guests in your street. Imagine your street now transformed into an "outdoor living room" with children playing, neighbours chatting and people dancing.' The organization LessTraffic.com provides useful guidelines on how liveable streets can be created. There are some existing street formats that do encourage this liveability: achieving such a vision is possible. Culs-de-sac can provide safe child spaces (see Figure 11.2), as can internal courtyard spaces (see Figure 11.3) and communal garden space where houses face green space, not roads. In terms of creating liveable streets where reduction of street traffic is addressed directly, the *woonerf* or Home Zone initiatives are instructive.

Imagine your street with 50 per cent less traffic – with children playing, neighbours chatting

The Dutch introduced the *woonerf* (living yard) concept during the 1970s, a concept that has since been replicated in other European cities. Streets are people focused, and socialization is encouraged by shared carriageways, reduced traffic speeds and prioritization of pedestrian use and supported by landscaping and traffic-calming measures. In the UK, the Home Zones initiative

Figure 11.2 Good neighbourhood design encourages active multi-age group play (Coleraine, Northern Ireland)

Source: Claire Freeman

began during the late 1990s with 14 pilot schemes, and has been given some support and recognition in legislation (Department for Transport, 2000) and by the release of *Home Zone Design Guidelines* by the Institute of Highway Engineers (IHE, 2002). Since these 14 early pilot projects, the Home Zone concept has taken off and been implemented with varying degrees of success across the UK (see Figure 10.6 in Chapter 10). While there are ongoing issues of traffic and street parking, a study by Tim Gill (2007, p22) in London, nonetheless, reports:

> Findings from completed schemes show consistent positive outcomes for children and for communities. Almost all schemes are popular and lead to a stronger sense of community, making it more likely that parents will feel happy about giving their children greater freedom outside the home as they grow up.

The Home Zones initiative mainly focuses on retrofitting existing streets, though its principles have been particularly effective in their application in new street design, following the principle of prioritizing pedestrians rather than cars, as in the example of Java Eiland.

Figure 11.3 Although not an especially appealing housing design, the courtyard is nonetheless well used and provides a safe place for meeting up and for play (Amsterdam, The Netherlands)

Source: Claire Freeman

Good Housing Design

There are some housing forms that encourage interaction, sociability and play, and others that actively discourage such activities. Less important than the individual house design, whether it is an apartment, single or three storey, terrace or detached, is the way in which the houses relate to each other. Good housing design meets the needs of children and their families for easy public interaction, but also provides privacy for when children or their families need to be alone. Successful housing can be assessed by reference to how well it addresses both its internal and external relationships. Table 11.3 summarizes some important housing relationships relating to children. It also poses questions that can be asked to assess the relationship. In a housing design there will be variation, especially with regard to internal spatial relationships; but the overarching principle of providing for both engagement and appropriate private space for children and their families is fundamental.

Good housing design needs to relate well to the local environment and follow principles of sustainable housing design. Sustainable housing works well for children by reducing environmental impact, providing healthy homes and creating homes that are socially, economically and culturally appropriate.

Table 11.3 Relationships to be considered in housing design

Different home forms and relationships	Example of questions that can be asked to assess the relationship
The relationship between homes in a geographic location	Do homes face each other to encourage interaction or away from each other to encourage privacy?
	Can a child access other homes yet still be within view of their own home?
The relationship between public and private space	Can all children in a housing complex use the internal play space or only those whose homes adjoin the play space?
The relationship between indoor and outdoor space	Does the internal space flow easily into an outdoor courtyard or green space?
The relationship between family privacy and communal interaction	Is the home surrounded by a tall fence or hedge, a low fence or hedge, or does it open onto unfenced communal land?
The relationship between adult and child access to internal spaces	Do children sleep in the same space as the rest of the family or do family members have individual spaces?
	Can children play in the family lounge/living room?

There has also been a substantial growth of interest in developing sustainable–green–eco housing. Guidelines have been developed for constructing sustainable homes. A key one is the UK government's Code for Sustainable Homes (Department for Communities and Local Government, 2008). The code scores homes according to the following criteria: energy and CO_2 emissions, pollution, water, health and well-being, materials, management, surface water runoff, ecology and waste. Another code of practice is the Leadership in Energy and Environmental Design (LEED) (Canada Green Building Council, undated) and the Green Building Rating System commonly used in the US, Canada, the UK and elsewhere. Key areas of focus in the LEED system include sustainable site development, water efficiency, energy efficiency, materials selection, and indoor environmental quality. Children benefit from homes that take into account geographic and climatic factors, as they are healthier and have more pleasant indoor and outdoor spaces. Good home design gives consideration to:

- access to sun and shade depending upon climate;
- optimizing solar bioclimatic principles to allow/preclude light penetration and reduce energy use;
- building with reference to aspect and topography in order to reduce wind exposure;
- natural ventilation providing better air quality;
- low energy consumption;
- energy-efficient heating;
- use of natural materials to reduce exposure to chemicals and toxins;
- durable design that can stand up to children's use;

- safe design to prevent falls and accidents;
- flexible design to meet children's needs at different ages;
- reduction of car space (garages, parking, roads) to enhance space available for other uses, especially play space.

Green design provides positive outcomes for children. Green roofs can enhance the area of available play space, especially in higher-density buildings. Water recycling and natural storm-water features can create water features such as ponds and streams for children's play. The use of ecologically appropriate planting supports and encourages the presence of wildlife in and around the home, providing opportunities for children to experience wildlife such as bats, dragonflies, deer, possums, foxes, insects and birds as part of their everyday lives. At present, ecological–green homes tend to be more expensive; however, this expense needs to be assessed against the negative costs associated with cheaper construction and ongoing costs such as higher heating costs, medical costs resulting from poor health, environmental costs coming from excessive water use, and opportunity costs resulting from loss of ability to play due to poorly sited or laid-out homes.

Child Spaces

There are places where attention to children's needs matters immensely as these spaces have children as their dominant user group. One such space is outdoor play space, which falls into four main groups:

1. public parks, usually multipurpose;
2. dedicated play areas, such as playgrounds;
3. institutional space, such as childcare centres, schools, sports facilities, BMX tracks and children's hospitals; and
4. spaces colonized by children informally, such as areas used by skaters, street-gathering points and places where families go to feed ducks.

Children inhabit a much wider range of places than formally recognized 'child' spaces

The problem with design, to date, has been that consideration has been given to children's design needs only in formally recognized 'child' places: those falling into the first three categories. Yet, children inhabit a much wider range of places than these formally recognized 'child' spaces and should be considered in all building and open space design where they will be current and future users of that space.

Here we identify five key considerations for good development:

1. *Careful site selection.* The first consideration in building a child-oriented space should be that the development site has been selected because it is the most appropriate site for that development. Too often the reverse is true, where children are allocated sites that are unwanted by other uses, are left over and/or are cheap. This explains the numbers of schools on the edges of suburbs that maximize the distances that children have to travel, or skate parks adjacent to heavy traffic roads or on derelict land rather than being in well-serviced sociable settings where skaters can access toilets, shops and eating places and which are close to public transport.

2 *Suitability.* The space should be designed specifically with the needs of the child user as the priority. Children too often end up in unsuitable spaces (especially buildings), usually because they were available, unwanted and cheap. Alternatively, they are in spaces normally used by adults where children are fitted in when adults are not using the space. Childcare buildings often fall into the unsuitable category. The drift away from community to privatized provision of childcare, where the emphasis is on profitability, has led to growing numbers of childcare facilities being in unsuitable buildings, originally designed for adult use, where access to good open space is especially compromised (Walsh, 2006).
3 *Quality.* The space should be designed with reference to quality design and materials, not made cheaply in a makeshift manner because it is for children.
4 *Access across a range of settings.* Child spaces should be considered across the range of spaces that children may use. One example is the city centre, where children's needs are rarely taken into account. Although children and their families are frequent users, city centres can lack convenient spaces for children and their families to relax and play. Another space is the hospital or hospice where children go as patients, but also as outpatients and visitors. Children's wards do go some way to meeting children's needs; but children can be frequent users of other hospital departments where they can spend long periods waiting for treatment, including Accident and Emergency, Haematology, Maternity and Obstetrics, Nutrition, Oncology and Radiology Departments. Again, they need spaces to relax and play while waiting. Visiting a hospitalized relative in a well-designed, sheltered, child-friendly courtyard space provides a far more positive experience than trying to 'contain' an inquisitive child in a public ward (see also Box 7.2 in Chapter 7).
5 *Choice.* There should be several spaces available to use, not one space. Multiple spaces are needed to meet the children's diverse needs. Playgrounds are a strong example of where diversity is needed.

Multiple spaces are needed to meet children's diverse needs

Parks and playgrounds have been the main focus of planning for play. Somewhat paradoxically, the number of parks and play spaces evident in a city can be seen as an indication of the degree of 'child unfriendliness' of a city. Ward (1978) refers to these dedicated spaces as child 'ghettos', arguing the child's right to play across the whole urban environment. Even so, the reality is that play spaces will continue to be designed for and perform an important role in children's play experience, and developers need to be guided in creating good children's space. Play space is especially important as a response to changes in urban living conditions in many cities:

> The problems inherent in suburban play space have become more pronounced in recent times. In today's suburbs, a children's play environment is quite different. Socially it tends to mean fewer siblings, carers spending longer hours in the workforce, higher parental expectations and a perception of public spaces as unsafe for children. Physically it translates into children being trapped inside built structures such as childcare centres, domestic dwellings and video arcades, or in supervised activities for a larger proportion of time. (Walsh, 2006, p138)

Although Walsh (2006) was writing from her position as a play consultant in Australia, the 'domestication' and movement inside of play is one reflected in many countries. Karsten (2005, p287) noted similar moves between past and present generations in Amsterdam, and in her interviews noted examples of 'outdoor play performed indoors', such as hide and seek and the building of huts (soccer was the most extreme example). This movement indoors was a response to the fact that, over time, public space has been transformed into adult space, with increasing concerns about safety and the reduction in free play time. Where outdoor play is most encouraged – namely, in designated play areas – safety and equipment provision have become the dominant drivers. Health and safety rules and the professionalization of play design have been strongly influential. Play space design has been captured by play equipment manufacturers whose uniform multicoloured designs colonize cities globally and whose constructions are seen as best meeting play safety standards. While safety is important, it can also frustrate attempts to design fun, challenging play spaces as the following example illustrates.

One of the authors was recently involved in a playground upgrade in New Zealand. The area is adjacent to one primary school, two high schools and is used by several local childcare facilities. Prior to the upgrade, the park comprised lawned space, trees, a small number of fixed play structures, swings, a climbing frame, slide and seesaw. When consulted by the city council at public meetings about the upgrade, many suggestions were made by members of the public on the type of improvements that they would like to see. Recognizing budgetary constraints, suggestions included using the superb Wellingtonia trees (very tall conifers) in the upgrade, perhaps with ropes for climbing or as supports for some equipment and providing structures that could function as both sitting and climbing spaces, particularly for the older high school students. As part of the consultation, catalogues showing equipment were circulated. However, given the fact that safety matting and design fees would be a major cost, the city council advised that any additional equipment purchases would be limited. It was also revealed that the local well-established play equipment manufacturer could not be used due to the fact that some aspect of their application for safety compliance to meet revised safety codes was still in process. It was stated that equipment from Denmark, used elsewhere in the city, might be imported instead (at much extra cost). In the end, a double slide, a swing set and a flying fox (very slow and aimed at young children) were added, together with some seats. What is most interesting is what wasn't included. There were concerns too much seating might attract too many young people, and dual-use structures on which children could play and sit could contravene safety requirements. Attendees at the public meeting were told that arborists would not like the trees used as play structures, ropes from the trees would be too dangerous, and use of the trees for play might require safety matting to be installed at the base of the trees. Nonetheless, use of the area has increased, indicating that even relatively small-scale improvements can markedly enhance use. This example is a story of lost opportunity, a story that dominates much play space provision. There are good principles available to use in designing good outdoor space for children, as indicated in Table 11.4. However, translation of these good principles into action requires appropriate resource provision, innovative thinking and commitment to providing not just safe but challenging and interesting play spaces.

A story of lost opportunity dominates much play space provision

Table 11.4 Design principles for children's outdoor space

Generic principle	Design principle	Design detail
Philosophy: the space fits with the programmes offered by organization. Child development: supports children's physical and social development needs. Access: at appropriate times, not just available to children when not otherwise in use. Ages and stages: provision for different needs; places for infants, disabled children, teenagers and adults, and the elderly who wish to be present. Dedicated space: included in the design, not as an afterthought. Ambience: feels warm, welcoming, friendly, open to use. Social space: places to watch, wait, mingle; large user group spaces and intimate spaces. Invitation: is it welcoming and does it invite use; does the entry point indicate safety, nurturing, transition and attraction? Choice: different spaces available in the locality, not just one space for all children.	Manipulation: children change the environment, build, dismantle, reorder, move things. Site: should visually relate to the external environment; all parts should be used, have variety and be environmentally safe. Safe access: accessible by public transport, support children's independent access, be car free. Size: should be able to accommodate the number of users. Sensory stimulation: sight and sound. Indoor–outdoor flow. Microclimate: protection from exposure, natural elements, drainage; provision for outdoor play in all weathers; uses local climate (e.g. sledding, water play). Quality and maintenance: high construction and maintenance standards; use of quality, durable and (ideally) natural materials.	Facilities: access to toilets, food, water. Variety and opportunity: range of activities and spaces, including quiet spaces and spaces for team games, wheeled toys, bikes, rollerblades and skateboards. Natural areas and planting: trees and plants for climbing, hiding, for wildlife, to provide shade and protection, to stimulate sensory perceptions. Child scale: furnishings; outlooks that children can reach, see, access independently. Activity paths and areas: well-defined areas, no dead space, separation of conflicting uses, open space. Equipment: play equipment; fixed structures such as swings and temporary use equipment such as balls, bats, rollerblades, scooters. Surfacing: range including natural surface grass and stones; hard surfaces for bikes and high use; soft surfaces for below equipment. Natural materials: water, sand, dirt, wood and building materials.

Is There an Ideal Built Form?

The ideal built form is not one that follows a particular set of building codes, but one that adopts general principles of good design for children. What matters are the relationships that the built form has with its child users. How well does the built space respond to the following questions: is it at a scale that children can engage with; is it welcoming and accessible; does it challenge and have variety and flexibility; does it promote vibrancy; does it incorporate natural features and sustainable resource use? The biggest challenge facing urban designers is retrofitting poor urban environments and making them work, as these are places where the most vulnerable children frequently live. There are some good models, or at least some good principles, contained in urban design models on how to design good child-oriented environments. However, these models (New Urbanism, green cities, liveable neighbourhoods) are mostly aimed at new developments. What the existence of these models shows is that there is no justification for the ongoing development of poor urban environments. Designers recognize the problems associated with shopping malls and leisure centres that are inaccessible to children other than by cars; the low-rise and 'McMansion' land developments with limited outdoor play space and designed to inhibit neighbourly interaction; and the continued upgrading and enlargement of roads in residential areas. The crux is to stop building them and to build better.

In his book *Child in the City* Colin Ward (1978, p204) put the case for a 'shared city', one where children share the city with all its users, not one where 'unwanted patches are set aside for children'. The key to good design is not merely the design of its constituent parts, houses, shops, roads and industries, etc., but its overall design. Too many children in Western society experience the city in a form that is described as 'islanding'. Each child has a series of islands that they journey between, almost like being parachuted into the spaces. Each child's use of the city becomes a response to an individualized and not a shared itinerary (Gillis, 2008). In part, this 'islanding' occurs as a result of children

Figure 11.4 Retrofitting urban areas to prioritize pedestrian rather than car accessibility benefits all ages and provides safer passage for children moving across the urban landscape (London)

Source: Claire Freeman

being pushed out from the increasingly adult 'mainland'. Good design is about reclaiming the mainland (the wider city), sharing the city space and creating child territories – that is, places where children share space and encounter other adults and children and move safely across this territory (see Figure 11.4). The *woonerf* and Home Zone models show how good design can achieve this goal and how streets can be reclaimed for people. For the suburbs, the liveable neighbourhoods concept offers good design principles. Built form matters; but there needs to be commitment on the part of urban designers to recognize children in urban design and to create buildings and spaces that support children's needs. For now, children still languish on the outer fringe of urban design consciousness.

Children still languish on the outer fringe of urban design consciousness

References

Appleyard, D. and Lintell, M. (1972) 'The environmental quality of city streets: The residents' viewpoint', *Journal of the American Planning Association*, vol 38, no 2 pp84–101

Birmingham City Council (2001) *Places for All*, Planning Department, Birmingham City Council, Birmingham

Canada Green Building Council (undated) *LEED Canada*, www.cagbc.org/leed/what/index.php

Childstreet (2005) *The Delft Manifesto on Child Friendly Urban Environment*, Childstreet Conference, 24–26 August 2005, Delft, The Netherlands, http://urban.nl/en/project/childstreet-2005

City of Los Angeles Department of City Planning (2009) *The 10 Principles of Urban Design*, www.urbandesignla.com/udprinciples.htm

de Botton, A. (2006) *The Perfect Home*, Channel 4, www.youtube.com/watch?v=yUgjaf1d8Fw

Department for Communities and Local Government (2008) *Code for Sustainable Homes*, Department for Communities and Local Government, www.planningportal.gov.uk/uploads/code_for_sustainable_homes_techguide.pdf

Department for Transport (2000) Transport Act 2000, DfT, London

Engwicht, D, (1999) *Street Reclaiming: Creating Liveable Streets and Vibrant Communities*, Pluto Press, Annandale, Australia

Francis, M. and Lorenzo, R. (2006) 'Children and city design: Proactive process and the "renewal" of childhood', in C. Spencer and M. Blades (eds) *Children and Their Environments: Learning, Using and Designing Spaces*, Cambridge University Press, Cambridge, pp217–237

Gill, T. (2007) *Can I Play Out ... ? Lessons from London Play's Home Zones Project Report*, London Play, London

Gillis, J. R. (2008) 'Epilogue: The islanding of children – Reshaping the mythical landscapes of childhood', in M. Gutman and N. de Coninck-Smith (eds) *Designing Modern Childhoods: History, Space and the Material Culture of Children*, Rutgers University Press, New Brunswick, NJ

Gleeson, B. (2006) 'Australia's toxic cities: Modernity's paradox', in B. Gleeson and N. Snipe (eds) *Creating Child Friendly Cities: Reinstating Kids in the City*, Routledge, London, pp33–48

Hart, J. (2008) *Driven to Excess: Impacts of Motor Vehicle Traffic on Residential Quality of Life in Bristol, UK*, MA thesis, School of Built and Natural Environment, University of the West of England, Bristol, UK

IHE (Institute of Highway Engineers) (2002) *Home Zone Design Guidelines*, IHE, London

Karsten, L. (2005) '"It all used to be better?" Different generations on continuity and change in urban children's daily use of space', *Children's Geographies*, vol 3, no 3, pp275–290

Lloyd-Jones, T. (2004) *Final Report of the Working Group on Urban Design for Sustainability to the European Union Expert Group on the Urban Environment*, http://ec.europa.eu/environment/urban/pdf/0404final_report.pdf

Local Governments for Sustainability (2010) International Council for Local Environmental Initiatives (ICLEI), www.iclei.org/

Malmö City Environment Department (2009) *Environmental Programme for the City of Malmö 2009–2020*, Malmö City Council, Malmö, www.malmo.se/sustainablecity

Malone, K. (2006) 'United Nations: A key player in a global movement for child friendly cities', in B. Gleeson and N. Snipe (eds) *Creating Child Friendly Cities: Reinstating Kids in the City*, Routledge, Oxon

McKenzie, E. (2005) 'Constructing the Pomerium in Las Vegas: A case study of emerging trends in American gated communities', *Housing Studies*, vol 20, no 2, pp187–203

Melbourne City Council (2006) *Towards a Better Public Melbourne: Draft Urban Design Strategy*, Melbourne City Council, Melbourne, Australia

Ministry for the Environment (2002) *People+Places+Spaces: A Design Guide for New Zealand*, MfE, Wellington, New Zealand

Newman, P., Beatley, T. and Boyer, H. (2009) *Resilient Cities: Responding to Peak Oil and Climate Change*, Island Press, Washington, DC

New South Wales Commission for Children and Young People (2009) *Built4Kids: A Good Practice Guide to Creating Child-Friendly Environments*, New South Wales Commission for Children and Young People, Sydney

New Urbanism (undated) *Principles of New Urbanism*, www.newurbanism.org/newurbanism/principles.html

ResilientCity.org (2010) www.resilientcity.org/

UNICEF (United Nations Children's Fund) (2009) *Child Friendly Cities: Fact Sheet, September 2009*, UNICEF National Committees and Country Offices, www.childfriendlycities.org

United Nations Human Settlement Programme (2009) *Planning Sustainable Cities: Global Report on Human Settlements 2009*, Earthscan, London

United Nations Population Fund (2007) *State of World Population*, www.unfpa.org/swp/2007/english/introduction.html, accessed April 2010

Walsh P. (2006) 'Reflections on what developers can do for urban children', in B. Gleeson and N. G. Sipe (eds) *Creating Child Friendly Cities: Reinstating Kids in the City*, Routledge, London

Ward, C. (1978) *Child in the City*, The Architectural Press, London

Ward, C. (1989) *Welcome Thinner City: Urban Survival in 1990s*, The Architectural Press, London

Weller, R. (2009) *Boomtown 2050: Scenarios for a Rapidly Growing City*, University of Western Australia Press, Claremont, Australia

World Bank (2005) *World Bank International Comparison Program Tables of Final Results*, February 2008, http://siteresources.worldbank.org/ICPINT/Resources/ICP_final-results.pdf

12

Professionals and Children: Working Together

> In the planning and designing of new communities, housing projects, and urban renewal, the planners, both private and public, need to give explicit consideration to the kind of world that is being created for the children who will be growing up in these settings. Particular attention should be given to the opportunities which the environment presents or precludes for involvement of children both older and younger than themselves. (Bronfenbrenner, 1972)

The Professional Influence

As communities grow they form hamlets, then villages, towns, cities and metropolises. As they move along this continuum, responsibility for and connection to the land and its development increasingly becomes less the preserve of individual citizens, and more the preserve of the land-use professional, guided by bylaws, policies, legislation, national standards and international conventions. Land-use decision-making becomes regulated by those with a vested interest in maintaining the quality of the environment; but also through the application of regulations and standards that not only aim at ensuring environmental quality, but which also support their own professional existence. Among the most influential in this regard are planners, architects, engineers, property developers, recreational and housing professionals and policy-makers, all of whom work together with a range of related professionals, which includes lawyers, educators, community developers and environmentalists. Between them these professionals can dictate the way in which the environment is shaped, its buildings, open spaces, the scale of development, accessibility, whether animals can be kept, the level of noise that people are allowed to make, whether fences can be erected, and even the colours of buildings and where children can play ball games. Their influence on children's lives is immense. Road-building decisions, for example, are pivotal in influencing how much traffic passes the child's front door or traverses the neighbourhood, or whether speed controls and bicycling lanes or safe crossing places are in place. All of this impacts directly upon children's play, social connectivity, access and independent mobility.

The influence of professionals on children's lives is immense

As the quote by Bronfenbrenner (1972) demonstrates, there is a long-standing awareness of the importance of the planning profession in creating cities for children. Yet, this awareness has not always translated into positive action by planners and other professionals towards creating better cities for children. This chapter identifies the different and often contradictory and competing roles of professionals in children's lives. It considers the need for professions to take into account not only the impacts of their actions and decisions upon children, but also their *modus operandi*, and how it can be switched from one that usually alienates children to one that is actively informed by children and their needs. In addressing the roles of the different professions, the current 'silo' way of working is challenged, arguing that an integrated interdisciplinary approach is the only way to effectively plan for children. For example, transport professionals need to work with housing development professionals and child health professionals on new housing developments.

The starting point for this chapter is the premise that children are capable social actors, with abilities to contribute to the improvement and development of the urban environments in which they live. The capacity for children to assist in improving the urban environment, to be involved in the planning process and to represent themselves and other children is often grossly underestimated by professionals. When professionals and children combine their efforts, the opportunities for creating better cities are immense. To realize these opportunities, however, demands that children and professionals treat each other with respect and the methods of engagement used must be appropriate and meaningful. Children must not be fobbed off with a 'box city' that doesn't influence the shape of the real city, or with creating an artwork for the cover of a city plan, yet having no input into its policies (see Figure 12.1). This doesn't mean that creative activities aren't appropriate, but that any activity must have a real influence on what happens, and nothing happens unless children are taken seriously.

Figure 12.1 An example of the box city approach: This approach works well when it feeds into wider planning decision-making

Source: Planning Department, Auckland University

A Participative Movement

In this chapter we focus on participation and use the following definition: 'the process of sharing in decisions which affect one's life and the life of the community in which one lives' (Lansdown, 1995, p17). This definition implies some degree of committed engagement with children, not just their tokenistic presence. Although community engagement, or public participation as it is usually termed in city planning, is now universally acknowledged as a 'good thing' and indicative of best practice, its genesis is quite recent. It was only in 1969 that Shelly Arnstein produced her still much-used ladder of participation, which demonstrates the continuum of participative engagement. The ladder was subsequently adapted by Hart (1997) to reflect children's participation. On the bottom rung of the ladder is 'manipulation'; above that is 'decoration' and 'tokenism', rising to 'child initiated shared decisions with adults' at the top of the ladder. Unfortunately, many participation initiatives with children have languished on the ladder's lower rungs. Typical examples of low-order 'participation' include dressing children in project t-shirts for a photo shoot, bringing children in to sing a song to open an event, or using child art to decorate some strategy or public document. In these cases 'participation' occurs without children being given opportunities to understand or meaningfully contribute to the development of the event or strategy. Higher-level initiatives, where children and adults work together on projects, are, however, becoming increasingly common. As the concept of participation takes off, and experience of participation develops, the higher-order participation methods should become more prevalent.

Low-order 'participation' includes dressing children in project t-shirts for a photo shoot

As participation and the ethos of community engagement has developed over time, it has become clear that society comprises many 'publics' and interest groups, and there is a need to recognize 'difference' in society. Initially, the focus has been on recognizing the needs of women, racial and ethnic minorities, and people with disabilities. Recognition of the needs of children and young people and their inclusion in participation has come much later. The catalyst for children's participation was the 1989 United Nations Convention on the Rights of the Child. The convention enunciated children's rights in a number of areas directly relevant to urban planning and development:

- Article 3: The right for actions taken concerning the child to be in their best interests;
- Article 12: The right to freedom of expression;
- Article 15: The right to meet others;
- Article 17: The right to access information;
- Article 29: The right to education for personal fulfilment and responsible citizenship;
- Article 31: The right to recreation and participation in cultural life and the arts (adapted from Office of the United Nations High Commission for Human Rights, 1989).

These rights were further strengthened through Local Agenda 21, which emerged from the United Nations Conference on Environment and Development (UNCED) held in Rio de Janeiro in 1992. Chapter 25 of Agenda 21, 'Children and youth in sustainable development', reconfirmed children's rights and stated that they should be given due consideration and included in matters relating to the environment. Ironically, the experience of children and young people at Rio itself was less than inclusive. The book *Rescue Mission Planet Earth* (Children of the World, 1994) contributed significantly to the Children's Agenda internationally.

The book itself, though, was a direct result of children's thwarted attempts to be heard at the Rio conference. Parallel to the main UNCED conference was a Youth Summit attended by 300 people from 97 countries. Promised a one-hour slot during the 14-day conference, the young people only got ten minutes of official coverage (attended by television cameras). The prime minister of Norway had promised to bring six world leaders to hear statements from the Voice of the Children's group, yet none of these leaders turned up. Their experience following Rio was better when around 10,000 children from 100 countries contributed to the Rescue Mission initiative, whose outcomes are the focus of *Rescue Mission Planet Earth* (Children of the World, 1994). Not being heard was, and sadly remains, a common experience for children.

The limited attention given to children at Rio was surprising given that the United Nations Convention on the Rights of the Child (UNCROC) was followed in 1990 (two years prior to Rio) by the United Nations World Summit for Children. It was attended by 159 governments, including 71 heads of state or government and 45 non-governmental organizations (NGOs). The summit agreed to 27 goals that the governments of the world would try to achieve for children, producing the *World Declaration and Plan of Action on the Survival, Protection and Development of Children* (United Nations General Assembly, 1990). A mid-term report on progress followed in 1996. Then, in May 2002, a special session (attended by some 700 participants) was held and the document *A World Fit for Children* was adopted by some 180 countries. The document reaffirms national and international commitment to providing for children's needs and states:

> The right of children, including adolescents, to express themselves freely must be respected and promoted and their views taken into account in all matters affecting them, the views of the child being given due weight in accordance with the age and maturity of the child. The energy and creativity of children and young people must be nurtured so that they can actively take part in shaping their environment, their societies and the world they will inherit. (United Nations General Assembly 2002, p7)

For children to contribute requires recognition of the multiple barriers faced by some children

National governments took it upon themselves to develop their own responses. Canada produced the document *A Canada Fit for Children* (Canadian Government, 2004), which confirmed children's right to participate as valued members of their communities in matters that affect them. It emphasizes that 'Children have much to contribute' but recognizes, too, that for children to contribute requires recognition of the 'multiple barriers faced by some children in Canada' (especially indigenous children) and there is a need to challenge the idea that one approach can fit everyone (Canadian Government, 2004, p19). Although the main concern in Canada's document is with children's welfare, matters pertinent to children's rights and needs in built environments are identified, including the right to affordable housing; to be part of safe, healthy and caring communities; not be exposed to environmental risks; to transportation strategies that encourage citizens to walk, bike and use public transport; and to recognize the concern and energy that children have for the environment.

Just as governments have responded since UNCROC to the need to recognize children in national policy, so too have children's needs and rights begun to be more generally acknowledged by academics, policy-makers and practitioners, ushering in an era of more positive commitment to engagement with children. A number of major initiatives and publications have provided strong support for children's participation in environmentally related matters and provided guidance on how children should be included in the participative process. The years 1997 to 1999 were particularly productive, with the release of a number of major publications on children's participation (Hart, 1997; Adams and Ingham, 1998; Freeman et al, 1999; Mullahey et al, 1999; followed later by Driskell, 2002; Chawla, 2002). Since then numerous local government child policy documents have been developed, to the extent that these are becoming the norm rather than the exception.

In city planning and development there is now international and national support for recognition and inclusion of children in urban planning and development. As signatories to international agendas and initiatives such as UNCROC and *A World Fit for Children*, governments have committed themselves to children's rights. The infrastructure for children's inclusion exists. The next section examines whether this translates into the day-to-day practice, policy and commitment of city land-use professionals.

Why Professions Should Work with Children

Professions matter because their decisions and practices affect children's lives. The tendency is to see children as being of interest to a small group of primarily medical and educational professions (doctors, teachers, and family and youth workers), where the focus is usually on matters of educational or social welfare interest. Yet, when looked at more closely, these professions can also have great influence on children's experience of their physical lived environments. Children's attendance at the local school rather than a more distant school was found to be the most significant factor contributing to children's independence and social connectedness in a neighbourhood study in New Zealand, and children attending local schools tended to be more active and independent in their neighbourhoods (Freeman, 2010). Thus, educational policies that encourage attendance at local schools can support children's local social networks and independence, including outdoor play. Doctors and environmental health workers can advocate on behalf of child patients in matters of poor housing provision and against the presence of noxious pollutants in the local environment.

Policies that encourage attendance at local schools can support children's independence

There are a number of ways in which professionals can consider children as partners in the environmental process and recognize the ways in which consideration for children's needs and rights can be justified and encouraged. One useful conceptualization is that of Francis and Lorenzo (2006), who identified what they called the seven realms of participation (see Table 12.1). The value of this approach is that the different professions can identify the realm or combination of realms that accord best with their own professional ethos and practice.

The Professional Record

The scope of professionals whose work impacts upon children's experiences of the urban environment is wide and includes lawyers, politicians, architects and engineers. Some of the key professions are indicated in Table 12.2, together

Table 12.1 The seven realms of children's participation in city design and planning

Realm	Approach
Romantic	Children as planners, children as futurists, planning by children.
Advocacy	Planners for children, adult planners advocating for children's needs.
Needs	Social science for children; research-based approach that addresses children's needs.
Learning	Children as learners; participation through education-related processes.
Rights	Children as citizens; protection of children's rights.
Institutionalization	Children as adults; planning by children within adult institutional boundaries/settings.
Proactive	Participation with vision; planning with children.

Source: Francis and Lorenzo (2006)

with the main areas where they currently consider children in their work practice. To date, the record of the professions in relation to considering children's needs and participative working with children is very low, with most – particularly developers, architects and traffic engineers – having no history of considering children's needs or working with them. Neither would most consider such work to be part of their professional remit. The impact of documents such as UNCROC and various national- and local-level child initiatives have made minimal impact upon the professional ways of working of many of the professions highlighted in Table 12.2. Yet it is incumbent upon these professions to consider children's interests in all developments. Children, like adults, are citizens and members of the community. Public bodies, organizations and professions have a duty of care towards their citizens. This is a moral obligation, but also increasingly a statutory obligation required through international conventions, national legislation, plans and policy documents.

There has been little direct response from the professions themselves, particularly land-use professions, on how they will both recognize children's needs in their professional activities and decisions, and involve children in the planning and development process. Searches of the policy documents and publications produced by key professional planning, engineering and architecture bodies reveals little attention given to children. One exception is the *Child-Friendly Communities Position Statement* by the Planning Institute of Australia (PIA, 2007). In this document, the PIA commits to endorsing children and their well-being as a planning priority. It identifies four challenges that need to be overcome in developing child-friendly communities: a lack of resources; the tendency to design children out of the built environment; lack of research and dissemination of information to built environment professionals; and narrow sectoralism. The statement identifies a number of positive ways in which planners will be encouraged to address children's needs and to work with children. Other professional institutes have child-related policies and initiatives, but they tend to be education focused, based on knowledge transfers rather than on meaningful co-working. The Royal Town Planning Institute (RTPI, 2007), for example, has an excellent manual for planners: *Education for Sustainable Development: Engaging with Young People*, which directs the reader to some excellent participative projects.

Table 12.2 Built environment and related professions whose activities impact upon children's lives

Professional area	Responsibilities	Main areas where children are considered	Participative working
Health	Physical and emotional health	Developing concern about living environments, obesity, respiratory disease, mental ailments (e.g. depression)	*Low*: focuses on effects on individual, not causes of poor health and well-being.
Planning	Land-use planning and development	Child-oriented facilities, schools, childcare centres, skate parks	*Low but developing*: starting to develop child participation approaches; but these are exceptions, not the norm.
Architecture	Buildings	Buildings with high child use (e.g. schools, childcare)	*Very low*: no noticeable trends towards working with children on building design, but are including children's needs more in design of child spaces.
Law	Law creation, implementation, negotiation	Protection of children's rights	*Low*: confined to working with individual children almost exclusively on matters relating to social welfare.
Housing	Housing provision, design and allocation	Provision of housing	*Low*: most focus around children within the shrinking state sector; provision in private sector largely related to supply and demand as determined by developers.
Education	Formal and informal education	Transfer of knowledge, educational opportunities, learning experiences	*Varied*: formal educational contexts usually adult led; informal tends to use more collaborative and action-based learning.
Recreation	Provision and management of recreational facilities, built and green	Play, health, environmental education, sports	*Low*: there are exceptions (e.g. working practices of some play and youth workers).
Transport/traffic engineers	Roads, public transport, access	Traffic calming, cycling and pedestrian routes	*Low–negative*: focus is on design, traffic management; many measures actively harm children.
Environment	Nature conservation, environmental initiatives, recycling	Environmental education, family-based nature access	*Medium*: directly work with children in environmental education and natural settings.
Developers	Building construction	Determine land use; site and building development	*None–low*: focus is usually on development for profit; some more enlightened developers do engage in some level of community engagement.
Community development and related welfare professions	Families and community groups	Family welfare focus, usually work with 'at risk' and marginalized children, families	*Medium*: welfare rather than environment focused; can be consultative but may be at individual child level.
Politicians and elected members	Decision-making community representation	Allocation of resources, budget; can advocate on children's behalf	*Low–medium*: focus is generally on adult voters, but some examples of participative initiatives, children's parliaments and youth councils.

Over the last ten or so years, innumerable and usually very good guides on how to work with children, often accompanied by relevant case studies, have been produced by organizations such as UNICEF (2010), Save the Children (2002), the National Children's Bureau (Kirby et al, 2003) and the New South Wales Commission for Children and Young People (2005). These guides are usually available online and provide links to relevant national legislation and policy, and can provide good guidance on local circumstances and issues (see Figure 12.2). The tendency has been to focus on youth participation; it should be noted that participation for the under-eights and preschool children has received very limited attention compared to that for older children and young people.

A particularly useful tool is the child impact assessment tool produced for the Children's Commissioner of New Zealand in conjunction with UNICEF. The tool is described as follows:

> Child impact assessment involves assessing a proposed policy, decision or activity to determine its likely impact on children. It can be seen as one way signatories to the United Nations Convention on the Rights of the Child (UNCROC) can fulfil their obligation under Article 3(1) to ensure that the best interests of the child become a primary consideration in all actions affecting children, including those undertaken by government bodies. (Mason and Hanna, 2009, p3)

This tool enables professionals and organizations to evaluate whether their policies and practices around children are working and making a difference.

A critical question in relation to the professions is whether the increased knowledge around children and the commitments undertaken by national and local governments have actually changed how professions work and made practical improvements for children. Unfortunately, this question is difficult to answer as governments are keen to trumpet new guidance and to publicize

Figure 12.2 Methods used need to be child appropriate: Unfortunately, in this example, while the intention was good, the children are too young to read the board

Source: University of Otago Planning Programme

child-oriented development (e.g. a new skate park); but it is less clear how representative these projects are of day-to-day practice.

A study looking at planning and children was undertaken to try and answer this question with reference to planners in New Zealand (Freeman et al, 2004). A total of 58 planners from 50 local authorities (representing 72 per cent of all New Zealand councils) took part in the study. Policy planners rather than regulatory planners were the prime responders, with a number of regulatory planners stating that they didn't feel the questionnaire, with its focus on children and young people, related to their work sufficiently for them to complete it. Regulatory planners take primary responsibility for considering the detailed management of the natural and built environment, and recommend approval of a whole range of developments that affect children's lives, from new buildings to determining allowable noise or industrial emission levels. Their low levels of understanding around how their work relates to children's lives is worrying.

Nearly half of the planners said that their councils do have strategies and policies relating to children and young people, though rarely were planners involved in their construction and rarely did they use them. Around one quarter of councils did involve children and young people in some way – perhaps a youth council survey, school activity or art project. Care has to be taken in how this 'involvement' is interpreted as, for some respondents, it sometimes refers to the fact that a general call for submissions was seen as including children and young people since, any member of the community could respond.

Planners want to work with children

Children were not a significant concern to planners working in local government. The presence of policy and guidance at national and local level made very little practical difference to planners' decision-making, and planners rarely worked with other sectors such as education or community development, where expertise with children was more likely to reside. The survey's findings were redeemed from gloom by the very positive attitude that planners had to working with children. They want to do it. When they had undertaken a project with children, it was as if a barrier had been crossed and their fears conquered, as the following quote from a planner about a skate park project indicates:

> The process has removed the fear and negative attitudes held by the community and the council towards the skateboarders. The success of the project has meant that the council is more likely to involve young people in the future in any similar projects. The enthusiasm of the young people who were involved helped create a positive attitude towards the project and gain the support required by both the councillors and council staff to push the project through. (Freeman and Aitken Rose, 2005b, p394)

Professionals, therefore, need to gain experience of working with children to build up relationships and, importantly, to identify appropriate resources and expertise within the community, organizations and government. Training on how to work with children should be part of professional education (see Figure 12.3). Working with children – more than any other group – demands inter-professional working and sharing of expertise.

Figure 12.3 For many years, working on joint projects with local school children formed an integral part of the education of planning students at Auckland University (New Zealand)

Source: Planning Department, Auckland University

Children's Participation

Children are especially likely to be disadvantaged with regard to participation in the community: 'Unlike most other marginalized groups ... children are not in a position within most western societies to enter into dialogue with adults about their environmental concerns ... In this sense children offer a special position of exclusion' (Matthews et al, 1999, p135). Similarly, Knowles-Yánez (2005) observes that nearly all land-use planning practices exclude children, even though there is a growing body of research that shows how and why they could and should participate. The exclusionary experiences of the children and young people at UNCED in 1992 could easily be repeated today, where children remain on the fringes of most environmental and developmental agendas. In policy terms, most governments now accept children's rights agendas, and professional bodies are seeking to move towards better engagement. Yet, progress is slow and children's participation is weak. When children are included, those who are do tend to be part of small, select, accessible groups with some 'hard to reach' children being repeatedly ignored. Those ignored include children who are different by virtue of culture, disability, geography, location or who are just simply perceived as 'difficult'.

The concept of 'difficult' does not necessarily accord with societal definitions of difficult as in problematic, troubled or at odds with society, but can refer to difficulties in terms of 'identifiability'. In fact, children and young people who are viewed as problematic, such as skateboarders and young people deemed 'at risk', are sometimes targeted in participation initiatives by council officers wishing to be seen as addressing a social problem. Some very inclusive participation initiatives have been undertaken with skateboarders around the development of skate parks. Skateboarders are an easily identifiable group;

they present problems (skating in public) where they are likely to come into conflict with other community members (e.g. the elderly and parents of young children) whose goodwill the council is unwilling to alienate (Freeman and Riordan, 2002). Another group whose participation is encouraged are the 'achievers'. These are children and young people most likely to volunteer, or to be identified as being able to contribute 'sensibly' to adult-initiated projects, and to take part in adult processes such as council meetings, preparing submissions or in parallel processes, such as youth councils and children's conferences. Thus, it can be the 'excluded middle' – the 'ordinary' children – who are often missed (Nairn et al, 2006).

The outcome of participation initiatives can be that the involved children unwittingly become victims of the system, as they find themselves in a position where they are seen as 'representing' the voices of an undefined mass of children rather than their own individual voices. The difficulties around representation have led to the disbanding of some youth councils where the 'representation' expectations of the children and young people sitting on the councils have been acknowledged as being unreasonable. Related to this is the need to acknowledge children's own individuality and to avoid the 'zodiac/star sign' approach to identity. The zodiac approach assigns everyone born within a certain period with identical characteristics and experiences – for example, the premise that all those born under the star sign Leo like power and all Librans are well balanced. Thus, skateboarders' views are seen as representative of the views of young people, although skateboarders represent just the views of one, usually male-dominated, sporting group.

Children involved in participation initiatives can unwittingly become victims

Councils have tended to work through groups that are identifiable and accessible, such as a youth group or a school class, and used the views of these children in a zodiac sign-type way as being indicative of all children of that age, from that place or with that interest. Children, like adults, represent the diversity of society. The photo of the children at the mosque is indicative (see Figure 8.4 in Chapter 8). While these children and others attending the same mosque are all Muslim, they are different. They have different ages, attend different schools, speak different languages, have different dress codes, and have lived in different countries; some are New Zealanders, while others are temporary visitors; they have different ways of practising their religion, live in different parts of the city and have different socio-economic status, all of which influence their views. Thus, even within this group it would be unwise to identify a 'Muslim child's view'. If this group were consulted, they would have varied responses; some children would be forthright, others quiet, some interested, others not, some would prefer written engagement, others something more active. Children are diverse: they are not a homogeneous group, just as a select adult group is neither homogeneous nor necessarily representative of wider social groupings. As Matthews and Tucker (2000, p309) state: 'There is no authentic voice of the child waiting to be discovered, only different versions of childhood ... multiple realities of children's lives and the many different cultural voices they represent.'

What's In It for Children?

The primary benefit of participation for children should be that the quality of their lives improves in some tangible way. This can be a practical benefit, as when

Table 12.3 The benefits of participation

Benefits for children	Participate in a new and exciting activity.
	Look at and understand their community and environment in new ways.
	Learn about democracy and tolerance.
	Develop new networks of friends, including community role models and resource people.
	Develop new skills and knowledge.
	Help to create positive change in the local environment and community.
	Develop a sense of environmental stewardship and civic responsibility.
	Develop confidence in their abilities to accomplish goals.
	Strengthen self-esteem, identity and sense of pride.
Benefits for other members of the community	Interact with children in positive constructive ways, overcoming misperceptions and intergenerational mistrust.
	Understand how children view the world, their community and themselves.
	Identify ways to improve the quality of life for young people.
	Build a stronger sense of community and pride of place.
	Appreciate the ideas and contributions of children.
	Invest time and energy in the future of the community.
Benefits for planners and policy-makers	More fully understand the needs and issues of the community.
	Make better and more informed planning and development decisions.
	Educate community members on the inherent complexities and trade-offs involved in policy and development decision-making.
	Implement, at the local level, the directives and spirit of the United Nations Convention on the Rights of the Child (UNCROC).
	Involve children in efforts to implement sustainable development.
	Create urban environments that are more child friendly and humane.
Practical benefits	More child-friendly environments.
	Safer communities – less conflict.
	Friendlier communities – more connections.
	More understanding and better-informed communities.
	Better built environment.
	More effective use of resources.
	Less conflict and time spent dealing with opposition to plans and developments.
	Quicker process (in the long term) of decision-making due to more engagement.
	Builds on and makes use of local knowledge and resources.

Source: adapted from Driskell (2002)

a new sports facility is built, or a less concrete but, nonetheless, equally important benefit, when children feel listened to and valued by their city council. If children benefit from participation, so do other members of the community and the professionals involved (see Table 12.3). These benefits can only be realized when a number of support conditions are met (Horelli, 2007). The process has to be undertaken in a supportive atmosphere. The children have to be comfortable with their selection for participation (children have the right to non-participation);

it needs to occur in a supportive physical space with encouraging mentors and information. However, if children have been properly prepared, the use of a high-powered decision-making space such as a council chamber can add authority to their voices, enabling them to be heard directly by city leaders. Children need to have access to the necessary information at an appropriate level so that they can understand it and contribute intelligently to the participative process. If they are participating in making a submission on a plan, they need to be aware of the significance of statutory plans, the timeframes for submission, the type of arguments that will be most influential to their cause, any financial or other possible resourcing issues that they need to consider, and so on. This is not to say that they should mirror in their submission what is often already an alienating process for many adults, but that they should not be penalized for their lack of knowledge. They need to maximize their chances of success and not have this reduced by being ill informed and ill prepared. It is thus important that the tools available to children are appropriate. Facilitating children's participation may require extensive changes to existing participative processes used for adults. This can be resource intensive. The final and most important condition is that children must be informed of the outcome of their participation. A poor participation experience can have lasting negative effects and 'turn children off' any further participation. Unfortunately, the UNCED experience at Rio discussed earlier is not unique and there is evidence of limited impact and disillusionment from those involved in participation initiatives (Hill et al, 2004).

Participation Methods

With participation, it is essential that the methods used are appropriate to the cultural context and to the children's age and interests. Fortunately, a range of possible methods are available, ranging from talking and listening, to resource-intensive technological methods (see Hart, 1997; Driskell, 2002). As there are many child participation guidelines and manuals available, we focus on a few issues primarily concerning visualization methods.

Great success has been achieved in urban projects with the inclusion of methods that enable children to directly present their interpretation, assessment and vision of the city and its development. Such methods include models, drawings, collages and photographic methods. Planners have been particularly keen on using these more visual methods; however, their use has not always been appropriate. Drawings or maps of places that children know, such as the local neighbourhood, are used to understand how children use, and what they value in, their neighbourhood (see Figures 12.4 and 12.5). They are appropriate and enjoyable child-centred activities, but can easily be misused. If children's map drawings alone are used to interpret children's spatial and social connections with the city, they can be highly misleading (Freeman and Vass, 2010). It is vital that in addition to producing drawings, children must also be interviewed at some length about their artwork for any meaningful information to be derived. These interviews should not be rushed, and we would argue that participation with children is necessarily time consuming and resource intensive. When challenged that this time commitment to children may be unrealistic for planners, we would argue that 'Just as it is no longer seen as appropriate to use one-off public

Great success has been achieved with methods that enable children to directly present their vision of the city

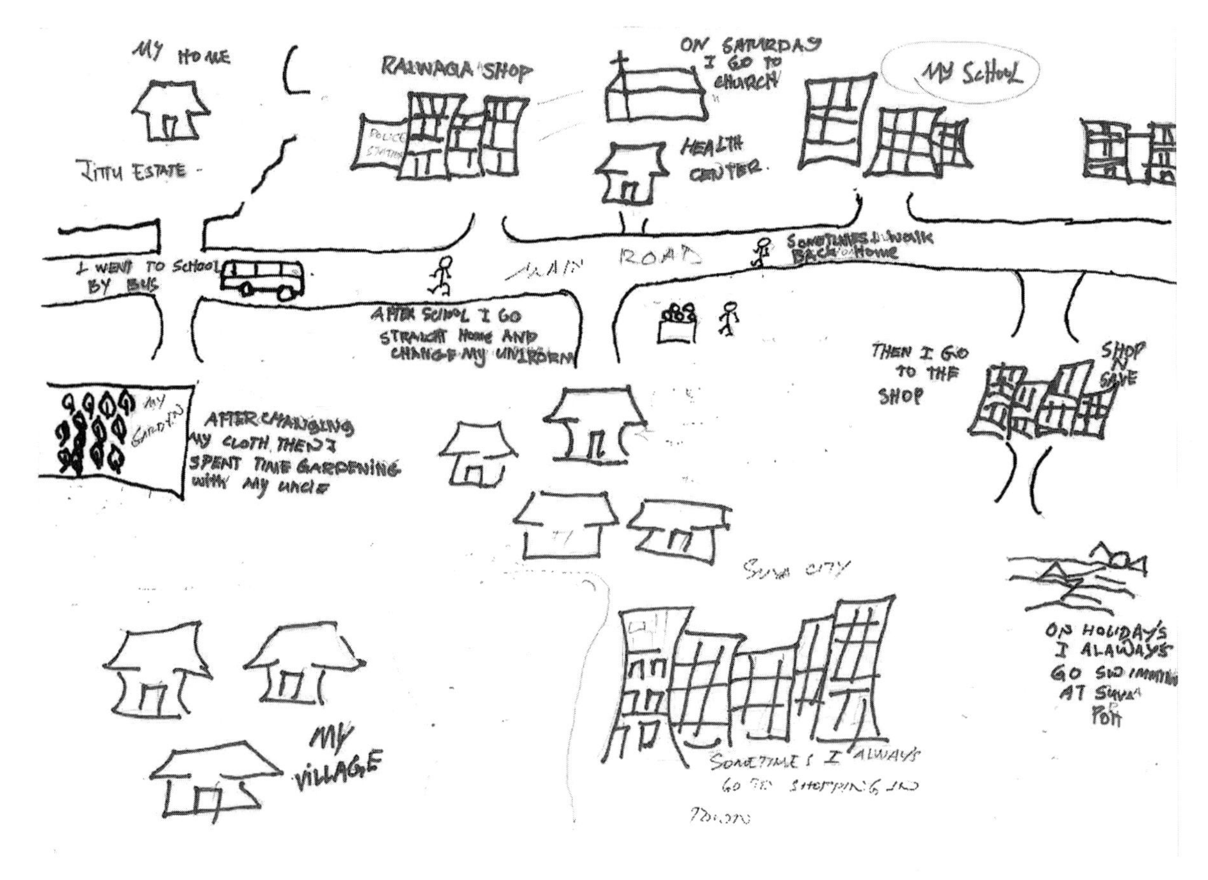

Figure 12.4 Maps can be used as a way of understanding children's life worlds, as indicated in this map from a child living in Suva, the capital of Fiji

Note: Similar maps have been used throughout the world.

Source: Boy from John Wesley Primary School, Suva, Fiji

meetings to gain community feedback on planning issues, so, too, a quick map-drawing exercise is inappropriate with children' (Freeman and Vass, 2010, p85).

Children are increasingly competent in their use of technology (see *Child Youth Environments*, 2009). This competence can be harnessed and used to facilitate and encourage more active and enjoyable participation. Photo exercises have been well used in participation and allow children to provide a direct record of their environment. However, they should be employed as part of wider discussion and dialogue, not as a record in their own right. Higher-order spatial visualization techniques such as computerized aerial photographs (Freeman, 2010), SoftGIS (Rantanen and Kahila, 2009) and Urban Mediator (Saad-Sulonen, 2008) are just three examples of technologies that can be used to facilitate participation. In the first example (aerial photographs), children aged 9 to 11 from very different socio-economic backgrounds and with varied experience of computers and air photo use all showed good competence in identifying spatial objects such as houses, roads and trees, and relating these to their personal experience. The technology enhanced their ability to demonstrate and share their spatial understanding of their neighbourhood and for some, also, the wider city.

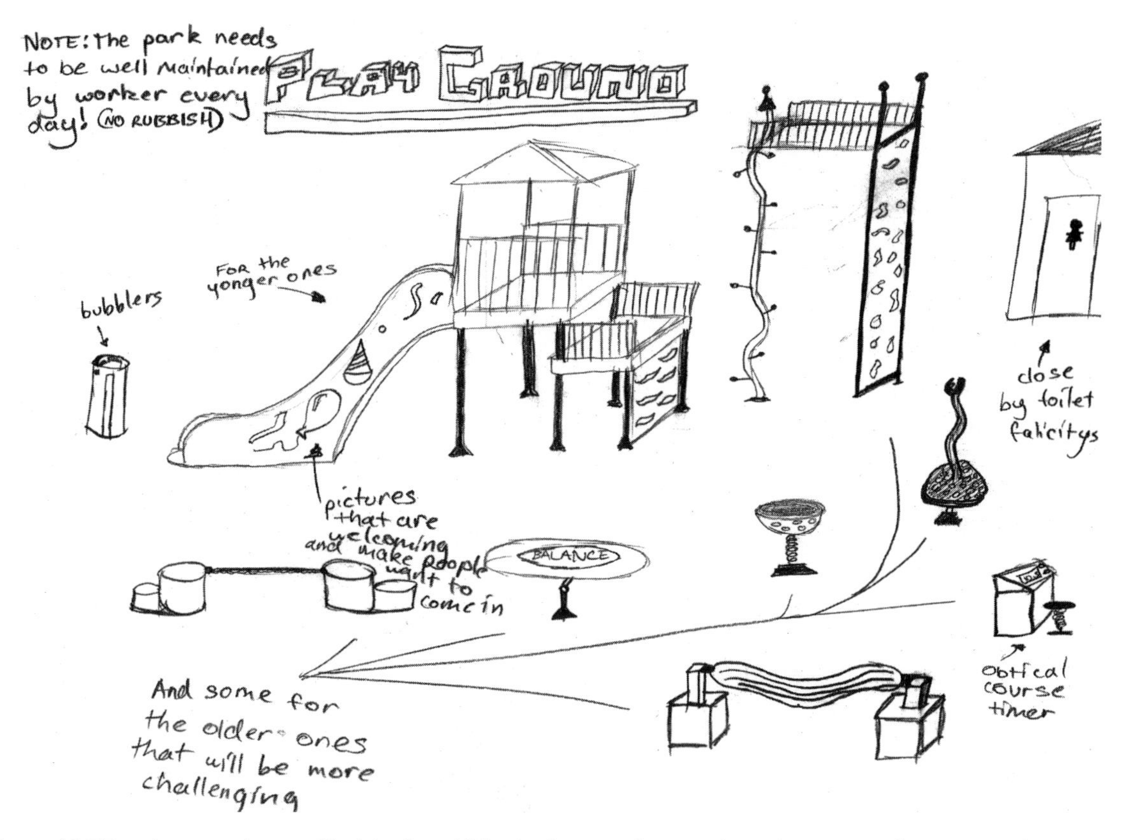

Figure 12.5 Drawings can give good insights into children's views on places such as play areas and can act as a focus point for discussion.

Source: Year 5/6 Child at Ngunnawal Primary School, Canberra

Computer games which involve the building of virtual cities, both historic (e.g. Caesar, Age of Empires) and contemporary (Cities XL) have proved popular with both children and adults, so much so that SimCity, the best-known city-building game, has been continually developed and re-released since its inception in 1989. Technology offers the potential to support the participation of larger numbers of children by using remote participation techniques. A geographic information system using a SoftGIS method to enable participation has been successfully used with children living in Turku, the third largest city in Finland (Kahila and Kytta, 2009). The SoftGIS method used in the study allows the children to identify places that are functionally, emotionally and socially meaningful; 11,000 such locations were identified. The most frequently mapped places were places to meet friends, to use the computer, to be at peace and to cycle. SoftGIS has also been used with 4000 adult inhabitants in six Finnish cities. The method provides valuable data that can be used by a range of professions, including planners, transport engineers, health professionals, youth workers, sports providers and council officers. Technology offers additional and engaging ways for children to engage in city planning and development. However, any technical method used should supplement – not replace – talking to children directly.

Understanding and Responding to Children's Needs through Participation

The selection of any one method (child conference, focus group, leaflet, children's day, children's workshop, children's council) will influence and determine the type, appropriateness and quality of the data gathered, as well as the child's participatory experience. Children can be included in, and contribute to, city development through a number of different access modes. Several of the most commonly used modes are indicated in Table 12.4. The access mode is important as the mode indicates the level of adult control and mediation. Thus, youth councils that mimic adult structures and occur within an adult political framework offer limited opportunity for child-initiated action and agendas compared to, for example, an internet-based discussion group, which can provide direct access for children to group members, discussions and information. Planners and professionals tend to work through established structures, such as schools, youth groups and youth councils, rather than contacting children present in informal gathering places, such as the park, playground, street, corner shop and bus stops. Accessing children informally is a valuable way of connecting with children who are not present or comfortable in formal structures and places. However, such contact is becoming increasingly difficult due to the presence of the (sometimes intense) ethical procedures, policy adherence, permissions and consent processes required for anyone seeking children's participation in an initiative, or even for talking to children in the public arena.

One very important area for participation is research. There have been significant shifts in focus in children's research from research *on* children to research *with* children, to research *by* children (Australian Research Alliance for Children and Youth and New South Wales Commission for Children and Young People, 2009). Consequently, children are being provided with more opportunities to become research participants rather than research subjects. Where children are research participants, the research is likely to be of greater interest and relevance to them, and to reflect issues that they, rather than the

Table 12.4 Access modes commonly used to facilitate children's participation

Access mode	Examples
Institutional structures designed for children	Youth councils, children's parliaments, school children's councils
Adult child representatives	Commissioner for children, ombudsman, child advocates
Project groups	Set up for a purpose (e.g. council reference group for new park development)
Discussion – informal groups	Web/internet-based group
Child/youth organizations	School, church, sport, interest groups
Political groups	Youth sectors (e.g Young Labour)
Place-based groups	Neighbourhood/community groups
City or environmental networks	Healthy cities networks, child-friendly cities networks, Mayors as Defenders of Children
Child/youth spaces	Schools, youth centres, informal gathering places

researchers, consider to be important. Where children are included, this enhances the research; however, not all research will be easy and not all research outcomes will benefit all participants. For example, in research on children's use of the city (Freeman, 2010), when discussing their maps and aerial photographs, some children could become aware of their own social isolation, as they may be unable to identify people in their neighbourhood and do not know who lives near them. The onus in this case is on the researcher to ensure that the children can make positive identifications on their maps and air photos, so that the child does not go away with a sense of failure derived from a 'socially' empty map. Not only is research with children necessary, but it needs to be in children's own interest. A think tank of experts on children's research met in Australia in 2008 and identified four key ways in which the research agenda needs to move forward to benefit children:

1. Build the capacity of children (integrate research within school curricula; train child researchers).
2. Build the knowledge of gatekeepers (e.g. ethics committees, council managers).
3. Build knowledge of researchers (provide guidelines; document and disseminate information).
4. Increase the influence of research (broad publishing, not just academic journals; target professionals) (adapted from Australian Research Alliance for Children and Youth and New South Wales Commission for Children and Young People, 2009).

Although the research agenda with children now includes some excellent guidelines and case study examples for urban professionals, especially those working in urban development, participation at this level is still poor. Where participation does occur, it is likely to be around specific projects, playground development, urban renewal, public art and tangible development schemes. What is still urgently needed is research that is at the starting point of the research continuum: research that informs professions of children's needs and what and how children's needs should be considered in urban planning. For example, research is needed on how children's needs should be considered in housing developments, city-centre shopping developments, and in public transport route development. While there is excellent research around children, much of this remains in the academic sphere or within the knowledge base of the project development team, and does not translate well into influencing professional practice more widely. Overcoming the gap between research and practice should be a priority.

Working for Children

Since the 1990s there has been tremendous progress on the part of professionals towards including children in environmental projects. When Roger Hart's book on participation came out in 1997 it was groundbreaking in its message of participation, working from the belief that 'only through direct participation can children develop a genuine appreciation of democracy and a sense of their own

Today, children's right to participation is widely acknowledged

competence and responsibility to participate' (Hart, 1997, p3). Today, children's right to participation, as well as the value of children's participation, is widely acknowledged at governmental level and also, in some cases, at the level of organizational responsibility (see earlier Planning Institute of Australia example). This recognition has yet to translate into effective process at the practitioner level, where planners, architects and traffic engineers rarely consider children in their developments; rarely work with children; and rarely consult guidelines, good practice manuals or child or youth policy documents or strategies. Where children do participate in urban development, the initiatives that they are involved in are sometimes cosmetic (creating artworks for plans but not being included in content discussions) or are one-off initiatives (inclusion in special one-off events, such as environmental clean-ups) which have limited long-term impact. Children and youth councils are one method used to overcome some of these limitations as they are longer-term initiatives, but are limited in their representativeness and by how seriously they are taken by the wider adult council.

Children's participation is likely to remain limited in the near future and focused on special initiatives and projects. Children, like adults, have the right to not participate, and professionals should recognize that their project may not be a priority for the children whom they seek to include. However, professionals are required under UNCROC and (depending upon the country's national legislation) national and local government policy to consider children in their practice. They do not have the right to ignore children's needs. It is in recognizing the need to take children into account in urban planning that the greatest change towards positive inclusion of children in city development could most effectively be made. To make progress, though, demands that professions take seriously their responsibility to consider children in all their development and land-use decision-making.

References

Adams, E. and Ingham, S. (1998) *Changing Places: Children's Participation in Environmental Planning*, Children's Society, London

Arnstein, S. (1969) 'A ladder of citizen participation', *Journal of the American Planning Association*, vol 35, no 4, pp216–224

Australian Research Alliance for Children and Youth and New South Wales Commission for Children and Young People (2009) *Involving Children and Young People in Research*, ARACY and NSW Commission for Children and Young People, Australia

Bronfenbrenner, U. (1972) *Two Worlds of Childhood*, Simon and Schuster, New York

Canadian Government (2004) *A Canada Fit for Children: Canada's Plan of Action in Response to the May 2002 United Nations Special Session on Children*, Canadian Government, Ottawa, Canada

Chawla, L. (2002) *Growing Up in an Urbanizing World*, Earthscan and UNESCO, London

Child Youth Environments (2009) vol 19, no 1, Special Edition on Children's Technological Environments

Children of the World in association with the United Nations (1994) *Rescue Mission Planet Earth Agenda 21*, Kingfisher Books, London

Driskell, D. (2002) *Creating Better Cities with Children and Youth: A Manual for Participation*, UNESCO and Earthscan, London

Francis, M. and Lorenzo, R. (2006) 'Children and city design: Proactive process and the "renewal" of childhood', in C. Spencer and M. Blades (eds) *Children and Their Environments*, Cambridge University Press, Cambridge, pp217–235

Freeman, C. (2010) 'Children's neighbourhoods, social centres to "terra incognita"', *Children's Geographies*, vol 8, no 2, pp157–176

Freeman, C. and Aitken Rose, E. (2005a) 'Future shapers: Children, young people and planning in New Zealand local government', *Environment and Planning C*, vol 23, no 2, pp227–246

Freeman, C. and Aitken Rose, E. (2005b) 'Voices of youth: Planning projects with children and young people in New Zealand local government', *Town Planning Review*, vol 76, no 3, pp287–312

Freeman, C. and Riordan, T. (2002) 'Locating skate parks: The planner's dilemma', *Planning Practice and Research*, vol 17, no 3, pp297–315

Freeman, C. and Vass, E. (2010) 'Maps and children's lives: A cautionary tale', *Planning Theory and Practice*, vol 11, pp65–88

Freeman, C., Henderson, P. and Kettle, J. (1999) *Planning with Children for Better Communities*, Policy Press, Bristol, UK

Freeman, C., Aitken Rose, E. and Johnston, R. (2004) *Generating the Future: The State of Local Government Planning for Children and Young People in New Zealand*, Department of Geography, University of Otago

Hart, R. (1997) *Children's Participation: The Theory and Practice of Involving Young Citizens in Community Development and Environmental Care*, UNICEF, London

Hill, M. D., Davis J., Prout, A. and Tisdall, K. (2004) 'Moving the participation agenda forward', *Children and Society*, vol 18, no 2, pp77–96

Horelli, L. (2007) 'Constructing a theoretical framework for environmental child-friendliness', *Children Youth Environments*, vol 17, no 4, pp267–292

Kahila, M. and Kytta, M. (2009) 'SoftGIS as a bridge-builder in collaborative urban planning', in S. Geertman and J. S. H. Stillwell (eds) *Planning Support Systems Best Practice and New Methods*, Springer, The Netherlands, pp389–411

Kirby, P., Lanyon, C., Cronin, K. and Sinclair, R. (2003) *Building a Culture of Participation: Involving Children and Young People in Policy, Service Planning, Delivery and Evaluation, Handbook*, National Children's Bureau, Department for Education and Skills, Nottingham and P. K. Research Consultancy, UK

Knowles-Yánez, K. L. (2005) 'Children's participation in planning processes', *Journal of Planning Literature*, vol 20, no 1, pp3–14

Lansdown, G. (1995) *Taking Part: Children's Participation in Decision Making*, Institute for Public Policy Research, London

Mason, N. and Hanna, K. (2009) *Undertaking Child Impact Assessments in Aotearoa New Zealand Local Authorities: Evidence, Practice, Ideas*, Institute of Public Policy, AUT University, Auckland, New Zealand

Matthews, H. and Tucker, F. (2000) 'Consulting children', *Journal of Geography in Higher Education*, vol 24, no 2, pp299–310

Matthews, H., Limb, M. and Taylor, M. (1999) 'Young people's participation and representation in society', *Geoforum*, vol 30, no 2, pp135–144

Mullahey, R., Susskind, Y. and Checkoway, B. (1999) *Youth Participation in Community Planning*, American Planning Association, Chicago, IL

Nairn, K., Sligo, J. and Freeman, C. (2006) 'Polarizing participation in local government: Which young people are included and excluded?', *Children, Youth and Environments*, vol 16, no 2, pp249–273

New South Wales Commission for Children and Young People (2005) *Participation: Count Me In! Involving Children and Young People in Research*, NSW Commission for Children and Young People, Sydney, Australia

Office of the United Nations High Commission for Human Rights (1989) *Convention on the Rights of the Child Adopted and Opened for Signature, Ratification and Accession by General Assembly Resolution 44/25 of 20 November 1989*, United Nations, Geneva, Switzerland

PIA (Planning Institute of Australia) (2007) *Planning Institute of Australia: Child-Friendly Communities Position Statement*, ACT, Australia

Ratanen, H. and Kahila, M. (2009) 'The SoftGIS approach to local knowledge', *Journal of Environmental Management*, vol 90, no 6, pp1981–1990

RTPI (Royal Town Planning Institute) (2007) *Education for Sustainable Development: Engaging with Young People*, London

Saad-Sulonen, J. (2008) 'Setting up a public participation project using the Urban Mediator tool: A case of collaboration between designers and city planners', in *Proceedings of the NordiCHI Conference 2008*, 20–22 October, Lund, Sweden, pp539–554

Save the Children (2002) *Children and Participation: Research, Monitoring and Evaluation with Children and Young People*, Save the Children, London

UNICEF (United Nations Children's Fund) (2010) 'Child and youth participation resource guide', www.unicef.org/adolescence/cypguide/resourceguide_planning.html

United Nations General Assembly (1990) *World Declaration on the Survival, Protection and Development of Children*, agreed to at the World Summit for Children, 30 September 1990, available at www.un-documents.net/wsc-dec.htm

United Nations General Assembly (2002) *A World Fit for Children: Resolution Adopted by the General Assembly, 11 October, Twenty-Seventh Special Session*, www.unicef.org/worldfitforchildren/files/A-RES-S27-2E.pdf

13
Conclusion: Children's Play and Resilient Cities

Now that many are beginning to awake to the danger, perhaps we will finally come to understand the stakes in the planetary emergency unfolding all around us – an emergency that involves far more than the pressing problem of climate change ... what confronts us is not a vague prospective danger to an abstract posterity in some future time. The threat is to a child born today. (Dumanoski, 2009, p1)

Children are a kind of indicator species. If we can build a successful city for children, we will have a successful city for all people, (Enrique Peñalosa, cited in Gilbert and O'Brien, 2005).

Looming Crises

The arguments in this book cannot be fully appreciated without understanding the global context of children's changing worlds. A recent surge of publications has highlighted the significance of a range of likely crises looming for humanity. There are stark descriptions of the probable consequences of a 'business-as-usual' model for our economies and societies (Rees, 2004; Brown, 2006; Leggett, 2006; Speth, 2008; Lovelock, 2009). Several feature film documentaries have helped to increase the public awareness of these issues – for example, Al Gore's *An Inconvenient Truth* (2006); *The 11th Hour* (2007), produced and narrated by Leonardo DiCaprio; and *The Age of Stupid* (2009), narrated by Pete Postlethwaite.

This book cannot be fully appreciated without understanding the global context of children's changing worlds

A central theme of these books and films is global warming, or global climate disruption, driven by the accelerating use and impacts of fossil fuels. Despite media coverage suggesting that scientific consensus estimates of climate disruption are exaggerated, recent research indicates that consensus assessments understate global climate disruptions, which 'may prove to be significantly *worse* than has been suggested' (Freudenburg and Muselli, 2010, p483). Global warming will 'destroy economies and eco-systems if more than a small fraction of remaining coal is burnt. Burning most of the remaining oil and gas will have the same effect' (Leggett, 2006, p198).

Climate change is not the only looming crisis. We can add several more, including an energy crisis in the form of 'peak oil' (the peaking and subsequent inevitable decline in global oil production); global food insecurity and mass refugee movements; growing international conflict over increasingly inaccessible resources; population overload; accelerating species extinction; changing distributions of disease and global pandemics; local environmental degradation; and global economic collapse (Meadows et al, 2004; Heinberg, 2007; Smil, 2008).

Fortunately, it's not all bad news for a child born today, and, as we argue below, children can help to develop creative responses to these challenges. Amid the warnings, there are messages of hope and suggestions for strategies that will remake our civilization to survive on a volatile Earth, as the title of Dianne Dumanoski's (2009) book implores. One of the most convincing arguments about the importance of these looming crises is provided by Thomas Homer-Dixon in his book with the intriguing title *The Upside of Down: Catastrophe, Creativity, and the Renewal of Civilization*. Homer-Dixon (2006, p17) acknowledges the difficulties of predicting the future, but also argues: 'we can still say confidently that we're sliding toward a planetary emergency'. Homer-Dixon recognizes the likely negative synergies that come from several crises occurring at the same time. He refers to the 'synchronous failure' that could occur when two or more of what he refers to as 'tectonic stresses' erupt.

Homer-Dixon's five tectonic stresses are:

- population stress from the rapid growth of megacities in poor countries;
- energy stress, particularly from the increasing scarcity of oil;
- environmental stress, evidenced in worsening damage to land, water, forests and fisheries;
- climate stress (global climate disruption); and
- economic stress from instabilities in the global economic system.

Homer-Dixon (2006, p12) argues that 'maybe not long from now' our social and technological systems may experience an 'abrupt breakdown'. The 'upside' is that this breakdown may then provide the catalyst that is needed to fundamentally change the way in which we operate as a society. He refers to this as 'catagenesis' – the process of breakdown, reorganization and renewal of a complex system. Such a catagenesis is more likely if we respond with a 'prospective mind' that engages positively with a world of great uncertainty and risk. Here, breakdown can be seen in a positive way: as a breakdown of the obstacles standing in the way of a change to new ways of doing things. The caveat is that if the collapse, when it occurs, is too severe, we will not recover from it. Hence, as Homer-Dixon argues, we need to build resilience in our economic and social systems.

We need to build resilience in our economic and social systems

Understanding the looming crises in our world is fundamentally important for children's future well-being (see Figure 13.1). More importantly, there may be opportunities for children that arise from these crises, including the opportunity for children to be involved as capable social actors in responding to the likely challenges. This may apply particularly to the issue of peak oil (Tranter and Sharpe, 2007).

Figure 13.1 The energy crisis is one that will be a significant part of children's future: here a boy powers a TV by pedal power at a community energy expo

Source: Claire Freeman

Child-Friendly Cities and Resilient Cities

Rather than be overwhelmed by the disquieting challenges of climate disruption, energy shortages, economic collapse and environmental degradation (among others), we argue that society should focus instead on the positive vision of creating child-friendly cities. In many ways, the characteristics of child-friendly cities are also the features that will make our cities more resilient in the face of several of these challenges to our economies and societies (Gilbert and O'Brien, 2005; Heft and Chawla, 2006; Tranter and Malone, 2008). Child-friendly cities are healthier, more liveable and more sustainable for all city residents.

Child-friendly cities are committed to fulfilling children's rights. These rights, as defined in the United Nations Convention of the Rights of the Child, can be grouped into three broad categories: protection rights, provision rights and participation rights. Related to these broad categories of rights, we can identify some key components of child-friendly cities. Children need to be protected from dangers such as traffic danger, assault, molestation, neglect and exploitation. Yet they should not be removed from all risk. Indeed, in our attempts to protect children from some risks, we have inadvertently increased their levels of inactivity and, hence, risks of obesity and related diseases, and the likelihood that children themselves will not develop resilience.

Children's provision rights are usually considered in terms of food, shelter, health and education. We should add to this the provision for playful exploration of their neighbourhoods, including the ability to manipulate their environments – to create their own play spaces, and also to connect with nature. Children's participation rights are yet to be fully appreciated by many policy-makers in cities. Participation includes not just formal involvement in local government decision-making, but also informal involvement in the social and cultural life of a community: a sense of connectedness with other people and a sense of place.

Children's rights may be compromised by a world affected by climate change, peak oil or population overload

Children's access to these rights may be compromised by a world affected by global climate change, peak oil or population overload. Yet, by concentrating *now* on creating child-friendly cities, the worst impacts of these forces may be avoided. Cities that meet the needs of children will not only create more resilient children, they will help to create more resilient cities.

The crux of a child-friendly city is the consideration of the best interests of the child in all actions and decisions taken by public or private institutions, as prescribed in Article 3 of the United Nations Convention on the Rights of the Child:

> If we were to take seriously the duty described in Article 3, it would be necessary to look at the implications for children when a new road was being proposed – what would be the implications for their health arising from noise pollution, lead pollution, poorer access to play facilities, decreased mobility and ability to move around within the local environment. (Lansdown, 1994, p40)

Such an undertaking would probably lead to a change from business-as-usual approaches, particularly for transport planning in cities.

The benefits of child-friendly cities in terms of creating resilient cities can be seen clearly in an examination of child-friendly transport. As explained in Chapter 10, child-friendly modes of transport include walking, cycling and public transport. Children use these independently, and adults using these modes make neighbourhoods safer for children, rather than more dangerous (as happens when cars are the dominant mode). These modes are also the modes advocated by researchers working towards the creation of more resilient cities, particularly in relation to the issue of oil vulnerability (Newman et al, 2009).

The transport systems of cities throughout the world depend upon access to cheap oil. Those that are least dependent upon oil are also less vulnerable to any threats to the supply of cheap oil. Not surprisingly, cities with superior public transport systems, and better provision for pedestrians and cyclists, use less oil. Citizens in these cities also have lower levels of obesity (Bassett et al, 2008). In the US, when cities are ranked by their ability to cope with an oil crisis, New York uses the least oil per capita (326 gallons per person) and Atlanta the most (782 gallons). In New York public transport accounts for 9 per cent of motorized trips, while in Atlanta it is less than 1 per cent (Newman et al, 2009). While this makes New York City appear more resilient, it falls well behind many European and wealthy Asian cities. In Munich, over 30 per cent of trips are made by public transport, and in Tokyo, 60 per cent.

Several cities throughout the world are increasing the mode share of public transport, walking and cycling. Not only in Copenhagen, but in Paris, Vancouver, New York and several other cities, innovative city governments and transport planning departments are replacing car space with space for people. New York, long known for its high public transport use (at least by US standards), is now making its streets more cycle and pedestrian friendly, with new cycle routes and pedestrian areas. In May 2009, New York Mayor Michael Bloomberg and Director of Transportation Janette Sadik-Khan embarked on the first stage of making parts of Manhattan's streets vehicle-free pedestrian plazas. Temporary barriers were erected in Times Square and Herald Square, blocking motorized transport. After a six-month trial, the scheme was deemed a success, and the barriers became permanent in 2010. Hundreds of metal tables and chairs are now a feature of what used to be a quagmire of traffic. The scheme is being extended to other streets, including Madison Square and Union Square, and both adults and children now enjoy these spaces (see Figure 13.2 overleaf).

The reclaiming of New York streets has arguably made the city more child friendly. It also helps to move the city away from a reliance on motorized transport, and, hence, increases resilience in terms of oil vulnerability. Encouraging the use of streets by pedestrians and cyclists (including children) also facilitates the spontaneous exchanges that make cities more enjoyable, liveable places.

Apart from child-friendly transport, other changes to cities to make them more child friendly can be seen to have benefits for making our cities more resilient in the face of coming challenges. We can see this for each of the activity spaces discussed in the chapters of this book: from the home, the school, the local neighbourhood, to the various service spaces, cultural spaces, natural spaces and the city centre. Some examples of how child-friendly spaces help to create resilient neighbourhoods and cities are outlined below.

Child-friendly spaces help to create resilient neighbourhoods and cities

In Chapter 3 we examined the features that make housing more child friendly. These features included safety from traffic and pollution, access to natural spaces and access to the wider environment. In terms of design, the yards of houses (particularly backyards, but also front yards), should be less concerned with aesthetics and more with allowing children to have contact with nature and to manipulate their environment (make dens or play equipment, or dig and play in the dirt). The trend towards high front fences or shutters closing off windows and doors should be avoided. The front yards and façades of houses should be designed to encourage, rather than inhibit, interaction with the street. Not only do these strategies make a more supportive home environment for children, they also have numerous benefits for creating resilient cities.

Figure 13.2 Playful children mixing with adults in a reclaimed Times Square, New York

Source: Paul Tranter

One important aspect of a resilient city is a strong locally based community, where people can support each other in the event of a crisis. This support could range from emotional support to practical support, such as helping neighbours to grow food. When homes are designed to encourage greater interaction between the houses and the street, this can help to maintain the contact needed to strengthen local communities. If a global economic collapse forces more people to grow food locally, this does not necessarily mean that children lose their play spaces – even in backyards. Children may find gardening a playful and rewarding experience, and gardens can provide play opportunities as well as contact with nature (e.g. using trees and vines as secret play spaces).

Similar arguments apply to the school. We argued in Chapter 4 that schools provide valuable opportunities for children to engage in creative play involving the manipulation of their environment and contact with nature, as well as opportunities for gardening and environmental learning. Yet few schools realize these opportunities. If we could, for example, encourage more use of schools for gardening, this might help the dissemination of valuable skills to cope with a possible food security challenge.

Some researchers have questioned our system of schooling, arguing that schools represent a flawed model for education in a world of looming challenges. Schools are currently places where children are taught how to become successful adults in a consumerist world, based on an economic system premised on a 'growth fetish' (Hamilton, 2004). Critics of this growth model, which views the Earth as exploitable for profit, argue that it is the fundamental basis for our current ecological crisis, and the continued promotion of this model causes the crisis to worsen. As one researcher asks: 'by educating students to fit into this system, are we not preparing them for a system that we know is broken?' (Steen, 2003, p200).

Are we preparing children for a system that we know is broken?

In an uncertain future, schools may need to become places where children are taught about self-sufficiency, conservation and cooperative activity. Schools could become important sources for local food production (vegetable gardens and fruit trees), and they may also become the foci for local neighbourhood community groups, for sharing ideas of food production, energy production and the creation of local employment or bartering systems.

For the 'neighbourhood' spaces of cities, some of the key aspects from a child-friendly perspective include the emphasis of the human scale (e.g. streets not highways and corner shops rather than large shopping malls), green spaces and spaces that encourage spontaneous exchanges. High levels of children's independent mobility are also an important part of good neighbourhoods. These features help with the development of resilient cities in a number of ways. Neighbourhoods where people can develop a sense of connection with each other – a local community – are likely to be healthier in every sense of the word. Social connections have been shown to be strongly related to health, and even to life expectancy (Seeman, 1996). The presence of children can also facilitate the strengthening of local neighbourhood-based community, as children are effective at breaking down the learned reserve between adults (see Figure 13.3 overleaf).

In addition to health benefits, stronger local neighbourhoods stimulate less motorized traffic and hence are less susceptible to oil vulnerability and create less greenhouse gas. If local neighbourhoods can generate local food production, this also leads to reduced food miles, as well as helping to generate stronger local networks.

For all the spaces discussed in this book, from the home to the city centre, a child-friendly strategy is likely to have significant 'co-benefits' (Smith and Haigler, 2008) for the health of all city residents, as well as the resilience of the city to future challenges, particularly peak oil and food security (Tranter and Sharpe, 2007). Not only will child-friendly spaces be more likely to be low-carbon spaces with greater use of local resources, they will also be spaces where children have greater opportunities to become capable social actors who can play a role in responding to the challenges described earlier in this chapter.

Figure 13.3 Breaking down barriers between children and those responsible for 'managing' urban environments is important for developing future relationships: young child talking to a guard in Washington, DC

Source: Paul Tranter

In order to create resilient cities, we first need to develop 'the capacity to respond with imagination and flexibility to a changing world' (Dumanoski, 2009, p9). Child-friendly cities have less segregation of children from other citizens, and when there is more opportunity for children to mix with adults, there is also more opportunity for an exchange of ideas and understandings. We should not underestimate the role or the capabilities of children. Children are more flexible in their thinking than are adults, and children are more open to new worlds and new possibilities. Because of this, they may well have a vital role to play in generating innovative strategies for the creation of resilient cities. As Rushkoff (1997, p13) argues, 'let's appreciate the natural adaptive skills demonstrated by our kids and look to them for answers to some of our own problems'. In other words, we should treat children as capable social actors in our quest to meet challenges such as climate change and peak oil.

If we do not involve children in the creation of resilient cities, we may simply lock ourselves into the mechanistic worldview that has led to the current reliance on technology, efficiency and complexity as a solution to our problems. If we theorize children not as vulnerable, dependent and incompetent, but as important, active social agents, then we can see children as part of the solution to emerging crises. If children are to become important social actors in a post-carbon world, they might be more involved in local food production, local composting and recycling, or repairing solar or wind-powered generating systems. Children will be more likely to walk or cycle to school and, hence, provide an important source of social contact for all local residents. They may have a role in selling local produce. Whatever roles they have, they should be encouraged to be involved in shaping our society and economy.

Conclusion: Changing Worlds

The last few decades have seen huge changes in children's lives in urban environments throughout the world. These changes have been driven by many interrelated forces, including the collective impacts of the decisions of parents, educators, bureaucrats, developers and politicians. When children have been considered in decisions, there have often been two important drivers: a desire to give children the best chance of becoming successful adults in a consumerist world, and the desire to protect children from harm – leading to a culture of risk aversion. These, and other forces, have led to the changes outlined in Chapter 2 of this book, especially the way in which children's lives are now more likely to be scheduled, indoors and with less contact with nature and more time in front of screens.

In the coming decades, we are likely to see continually changing worlds for children. The changes may be driven by broader global changes, and our responses to these changes. Parents may be less concerned about children being injured playing in playgrounds than they are about the impacts of a collapsed global economy, food insecurity or the impacts of international conflict.

Children's lives are influenced by both the characteristics of the spaces that they occupy, and by the social environments that direct the use of these spaces. A child-friendly environment is one where spaces support children's rights, particularly the right to play. Children also have a role in determining both the physical and the social spaces of their cities.

The primary business of children is play. Play is the way in which children learn about themselves and their place in the world. It is intrinsically motivated, spontaneous, directed by children rather than adults, and can occur in any context. Children can and do play anywhere and everywhere. When researchers who focus on children's environments understand the importance of play, then it is clear that an important goal of any planning policy or decision regarding children should be to facilitate play, particularly play involving some contact with nature. The concept of 'nature-deficit disorder' has been identified as a problem for modern children – children spending less time outdoors, resulting in a range of behavioural problems (Louv, 2005). The concept has struck a chord with both environmentalists and children's advocates, and there is growing evidence of the educational and health benefits of contact with nature (Maller et al, 2006). Research also shows that values supporting ecological sustainability grow out of children's regular contact with, and play in, the natural world (Moore and Wong, 1997). But perhaps there is a more fundamental deficit that needs to be addressed: *play-deficit disorder*.

Play is the way in which children learn about themselves and their place in the world

Figure 13.4 Alternative uses for cars in the city: Car used as a garden in Kensington Market (Toronto, Canada)

Source: Nick Tranter

As well as a lack of contact with nature, children's lives in modern urban environments are less playful than they were a generation or more ago. Children are now more likely to be engaged in adult-directed activities such as sport, music or extra-curricular academic coaching. These are not 'play', and moreover, children are often dependent upon adults to drive them to these activities.

If we can create cities in which children are freer to playfully explore their environments, then we are likely to have overcome many of the obstacles to the creation of resilient cities. Our transport systems will favour active modes. Shops, schools and services such as post offices will be closer to children's homes. Local communities will be stronger. We will find other uses for cars (see Figure 13.4), traffic speeds will be lower, and our lives may be at a more relaxed pace. Parents will have more time to spend on leisurely pursuits such as gardening, as their 'free-range' children will require less supervision and transport.

A logical conclusion of this argument is that we need not be paralysed with fear when considering the challenges ahead. This is not to deny the reality of climate change, oil vulnerability or global economic instabilities. Instead, it is to encourage a more confident, optimistic response, and a positive vision of our future cities. It is to take a lead from children, to experiment, take risks and explore the unknown with anticipation and excitement (see Figure 13.5). Creating resilient cities? It's child's play.

Figure 13.5 We can build on children's desire to take a risk with play and to take different paths in building better futures: children ignore the main bridge, preferring the challenge of the log crossing

Source: Claire Freeman

References

Bassett Jr, D., Pucher, J., Buehler, R., Thompson, D. and Crouter, S. (2008) 'Walking, cycling, and obesity rates in Europe, North America, and Australia', *Journal of Physical Activity and Health*, vol 5, no 6, pp795–814

Brown, L. (2006) *Plan B 2.0: Rescuing a Planet Under Stress and a Civilization in Trouble*, Earth Policy Institute, Washington, DC

Dumanoski, D. (2009) *The End of the Long Summer: Why We Must Remake Our Civilization to Survive on a Volatile Earth*, Crown Publishers, New York, NY

Freudenburg, W. and Muselli, V. (2010) 'Global warming estimates, media expectations, and the asymmetry of scientific challenge', *Global Environmental Change*, vol 20, no 3, pp483–491

Gilbert, R. and O'Brien, C. (2005) *Child- and Youth-Friendly Land-Use and Transport Planning Guidelines*, Centre for Sustainable Transportation, Toronto, Canada

Hamilton, C. (2004) *Growth Fetish*, Allen and Unwin, Sydney, Australia

Heft, H. and Chawla, L. (2006) 'Children as agents in sustainable development: The ecology of competence', in C. Spencer and M. Blades (eds) *Children and Their Environments: Learning, Using and Designing Spaces*, Cambridge University Press, Cambridge

Heinberg, R. (2007) *Peak Everything: Waking Up to the Century of Declines*, New Society Publishers, Gabriola, British Columbia, Canada

Homer-Dixon, T. (2006) *The Upside of Down: Catastrophe, Creativity, and the Renewal of Civilization*, Knopf, Toronto, Canada

Lansdown, G. (1994) 'Children's rights', in B. Mayall (ed) *Children's Childhoods: Observed and Experienced*, Falmer Press, London, pp33–44

Leggett, J. (2006) *Half Gone: Oil, Gas, Hot Air and the Global Energy Crisis*, Portobello, London

Louv, R. (2005) *Last Child in the Woods: Saving Our Children from Nature-Deficit Disorder*, Algonquin, Chapel Hill

Lovelock, J. (2009) *The Vanishing Face of Gaia: A Final Warning*, Basic Books, New York, NY

Maller, C., Townsend, M., Pryor, A., Brown, P. and St Leger, L. (2006) 'Healthy nature, healthy people: Contact with nature as an upstream health promotion intervention for populations', *Health Promotion International*, vol 21, no 1, p45–54

Meadows, D., Randers, J. and Meadows, D. (2004) *The Limits to Growth: The 30-Year Update*, Chelsea Green, White River Junction, VT

Moore, R. and Wong, H. (1997) *Natural Learning: The Life History of an Environmental Schoolyard: Creating Environments for Rediscovering Nature's Way of Teaching*, MIG Communications, Berkeley, CA

Newman, P., Beatley, T. and Boyer, H. (2009) *Resilient Cities: Responding to Peak Oil and Climate Change*, Island Press, Washington, DC

Rees, M. (2004) *Our Final Hour: A Scientist's Warning – How Terror, Error, and Environmental Disaster Threaten Humankind's Future in This Century on Earth and Beyond*, Basic books, New York, NY

Rushkoff, D. (1997) *Children of Chaos: Surviving the End of the World as we Know It*, Flamingo, London

Seeman, T. (1996) 'Social ties and health: The benefits of social integration', *Annals of Epidemiology*, vol 6, no 5, pp442–451

Smil, V. (2008) *Global Catastrophes and Trends: The Next 50 Years*, MIT Press, Cambridge, MA

Smith, K. and Haigler, E. (2008) 'Co-benefits of climate mitigation and health protection in energy systems: Scoping methods', *Annual Review of Public Health*, vol 29, pp11–25

Speth, J. (2008) *The Bridge at the End of the World*, Yale University Press, New Haven, CT

Steen, S. (2003) 'Bastions of mechanism, castles built on sand: A critique of schooling from an ecological perspective', *Canadian Journal of Environmental Education*, vol 8, spring, pp191–203

Tranter, P. and Malone, K. (2008) 'Out of bounds: Insights from Australian children to support sustainable cities', *Encounter: Education for Meaning and Social Justice*, vol 21, no 4, pp20–26

Tranter, P. J. and Sharpe, S. (2007) 'Children and peak oil: An opportunity in crisis', *International Journal of Children's Rights*, vol 15, no 1, pp181–197

Index

academic performance as priority 63–64
action research 65
Adelaide 103
adult-dependent mobility (ADM) 11, 23, 181, 187–189
  cost and time spent 183–184
adults
  fewer available for childcare 45
  negative impacts of change on 34
  rules 51–52
  see also adult-dependent mobility (ADM)
adult-structured activities 24, 165
adventure
  neighbourhoods 80
  playgrounds 120–121
  with public art 104–105
aerial photographs 236
affordances 31–32, 34, 64
aggressive behaviour 61–62
Aitken, S. 45
Amsterdam 142–143, 209
  Java Eiland 210–211
  right to the city 12
  time spent in the home 47–48
amusement arcades 126
Armstrong, D. 122
Arnstein, Shelly 225
Asia Pacific Child Friendly Cities Network 12
Attenborough, David 159
Austin, Texas 170
Australia
  Child-Friendly Communities Position Statement 228
  children's views on museums 131–132
  children's views on school grounds 71–72
  Family Law Amendment (Shared Parental Responsibility) Act 46
  housing developments 21–22, 107
  independent mobility (CIM) 185–188
  Liveable Neighbourhoods Initiative 208–209
  magpies 169
  multiculturalism 140
  play opportunities in schools 58
  public transport 189–190
  research with children 239
  TVs in children's bedrooms 50
  see also individual cities

backyards 48, 169–170, 247
Barbour, A. C. 65
Barker, J. 187
Bartlett, S. et al. 77
bedrooms, TV in 50
Belfast 103–104
  City Council 96–97
Bellamy, David 159
Benesse Educational Research Centre, Tokyo 24
Berkeley, California 65–66
Berlin 190
Betsky, Aaron 211
Birmingham 103, 142
blended families 14, 45
Boca-Baraccas, Buenos Aires 29
Bologna, Children's Manifesto 205
box city approach 224
Brazil 204
Brisbane 104–105, 170–171
Bronfenbrenner, U. 223

Canaansland, South Africa 82
Canada
  commitment to children's needs and rights 226
  community self-assessment 80
  independent mobility (CIM) 186
  multiculturalism 140
  working hours 48
car ownership 184, 193
Carson, Rachel 160
celebrations and festivals 79, 151–154
centralization of public services 115, 127
Chawla, L. 20, 29
Chawla, L. and Malone, K. 20
Chicago 175
child abuse and poor housing 44
childcare settings 48
child-free developments 107, 211–212

child-friendly cities
  literature 12
  and resilience 245–250
childhood 3–5
  ghettoization of 117–118, 217
  history of 7
  see also conceptualization of childhood
child impact assessment 230
child policy documents 226–228
child poverty 43–44
children
  capabilities of 129, 224, 250
  marginalization of views and needs 20
  use of space 9–10, 15
  views of the world 20–21, 31–32
  visibility of 31, 93, 211
Children and Nature Network 161
city centres
  20th century developments 95–96
  children's views on 100–101, 111
  child spaces 217
  conflicting uses 100
  essential qualities 93, 95, 112
  Federation Square, Melbourne 101–102
  functions 94–95
  as home 106–108
  multiculturalism 142
  public art 98–99, 101–106
  public space 97–100
  regeneration 96–97, 108, 204–205
Clements, R. 24, 31
climate change 16, 209, 243–244
codes of practice 215
cognitive development 63
Coin Street Community Builders 109
Coleman, J. S. 78
collective responsibility for children 27, 196
  Germany and Japan 30, 190
  see also passive surveillance
community 77, 248
  changes in structure 23
  development initiatives 79–80
  importance of socialization 87, 89, 249
  see also participation
community gardens 34, 122–123
computer games 237
computers 51
conceptualization of childhood 7–9, 198
  adults-in-training 57
  as dangerous 123–124
  impact on spatial design 116
  in museums and libraries 128–129
  as vulnerable 60, 192
conkers 169
consumer culture 152
Cooper Marcus, C. and Francis, C. 100
Cooper Marcus, C. and Sarkissian, W. 40–42, 54
Copenhagen 103, 193–194
cultural differences 30–31, 141–142, 155
  connections with home country 47
  family size 45
  independent mobility (CIM) 190–191
  Muslim populations 147
  see also segregation and inclusion
cultural life *see* public art
culture 137–138
  celebrations and festivals 79, 151–154
  changing 196–197
  children's right to 139–140
  internationalization of 152, 154–155
  multiculturalism 140–141
Cunningham, C. and Jones, M. A. 118
Curitiba, Brazil 193–194, 204
cyber-bullying 28
cycling, Copenhagen 193

deaf children 146
de Botton, Alain 211
decentralization of city centre services 96
Denmark 51, 190, 193
design *see* planning and design; urban design
disabled children 4, 118–120, 145–146
Disneyfication 152
divorce and separation, living in two homes 45–47
domestication of play 10
Donahoo, D. 7
drowning 27
Dublin 124
Dudrah, R. K. 142
Dunedin
  children's views on classrooms 71–73
  children's views on homes 53–54
  children's views on neighbourhoods 89–90

independent mobility 30
Durkheim, E. 15
Dyment, J. E. and Bell, A. C. 67–68
dynamic nature of space 5

Edinburgh urban regeneration
area 29–30
electronic surveillance 27–29
Elsley, S. 29–30
endangered species 161
energy crisis 16, 193, 243–244, 246
Engwicht, D. 196, 212
environment
disadvantaged areas 29–30
education 64–66, 68, 73
green city movement 86
threats to 16, 160–161
see also natural world
environmental amnesia 163
environmental relationships, emerging
literature 9–12
ethnicity *see* cultural differences; culture
exclusion *see* segregation and inclusion

families 14, 45, 84
activities 122, 164
size 23, 48, 143
fear
city centres 100–101
culture of 24–27
geographies of 11–12
of misconstrued motives 190
and restrictions on access to
nature 169
segregation and conflict 150–151
Fiji 82
Finland
independent mobility (CIM) 30, 168,
186–187
SoftGIS 237
food production 248–249
forest schools 68
Francis, M. and Lorenzo, R. 227–228
freedom
loss of 19, 23
and mobile phones 28
see also independent mobility (CIM)
Freeman, C. 30, 89
Freeman, C. and Aitken-Rose, E. 6–7
Freeman, C. et al. 231
frog life-cycle 162–163
front porches 191
Furedi, F. 7–8, 79

games 152–153
garden cities 83
gardening 122, 164, 249
gated communities 107, 211–212
Gateshead 25
gender
bias in school sport and games 67
cultural expectations 141
fear of victimization 25
girls' access to playgrounds 143
Germany
collective responsibility 30, 190
independent mobility (CIM) 185
public transport 189
ghettos 142
Gilbert, Oliver 175
Gill, T.
fear of misconstruction 190
Home Zones 213
neighbourhoods 78, 211
risk-averse behaviour 8, 68
undesirable activities 169
Gleeson, B. 117, 203
globalization
culture 152, 154–155
increased competition 191
Globe Sustainable City awards
204–205
Goodall, Jane 159
Good Friday Agreement 150
good parents 25, 196–197
Google Earth 173
Green Building Rating System 215
green cities 172–175, 209
green roofs 175, 216
Growing Up in Cities research
initiative 20

Habitat for Humanity 43–44
hanging out with friends 116, 123–124
Hardin, Garrett 160
Harker, L. 43
Hart, R. 10, 118, 172, 212
participation 239–240
health 15–16, 26, 126–128
fostered by good
neighbourhoods 80–82, 86–87
importance of natural world
164–165
indigenous children 149
see also obesity and diabetes
health and safety in playgrounds 218
helicopter parenting 23

high-rise housing 40–41, 83–85, 107
Hillman, M. et al. 30, 185, 187, 190
history and identity 103
Holloway, S. and Valentine, G. 51
home-alone 49
Homer-Dixon, T. 244
homes
  changing families 45
  children moving between two 46–47
  children's views on 53–54
  green spaces accessible from 173–175
  immigrants' home country 47
  importance of 39
  nature and variety 39–40
  needs of children 40–43
  ownership 83
  poor quality housing 43–44
  rules 51–52
  technology space 49–51
  time spent at home 47–49
  see also housing
homework space 50
Home Zones 194–195, 212–213
hospitals and healthcare services 94–95, 126–128, 217
Hough, M. 172
housing
  affordability 207–208, 216
  backyards 48, 169–170, 247
  children's needs 40–43, 247
  design 52–53, 84–87, 191, 214–216
  developments 21–22, 31, 40, 83–84, 210–212
  inner cities 106–108
  poor and overcrowded 43–44, 165
  see also homes
Howard, Ebenezer 83, 89

imaginative play 24, 69, 119
immigrant children 47, 84
inclusion *see* participation
independent mobility (CIM)
  decline in 5–6, 11, 23, 168
  environmental importance 184–185
  and housing density 31
  importance for children 182–183
  international comparisons 185–187
  reasons for variability 189–193
  recent trends 187–189
  reversing trends 197–198
  rural and urban areas 30–31
  and socio-economic status 30
  strategies to increase 193–197
  undermined by mobile phone 29
  value for neighbourhoods 185
  value for parents 183–184
indigenous children 149–150
individualism 21, 26–27, 30, 117
interviewing children 235
islanding 220–221

Jacobs, Jane 96
Janko, Susan 59
Japan
  celebrations and festivals 79
  game centres 126
  good neighbourhoods 86
  homework space 50
  independent mobility (CIM) 186–187
  normalization of indoor play 24
  public transport 189, 246
  risk-taking 62
  school ground greening 68
Joerchel, A. C. 141

Karsten, Lia 47–48, 138, 142–143, 218
Kids in Museums 129–130
Kotler, J. A. et al. 50
Kunstler, James 133
Kytta, M. 30

land-use decision-making 223, 232
law courts 95
Leach, Penelope 3
Leadership in Energy and Environmental Design (LEED) 215
learning
  environmental 64–66
  from nature 162–163, 169
  through play 63
  leftover spaces 100–101
  legal liability for accidents 60–61
Lerner, Jaime 194
Lets Win Back Our Cities congress, WWF 205
Liberty Swings 120–121
libraries 128–129
Liverpool 107
living statues 103
local spaces 120
London
  child poverty 44
  Coin Street development 108–110
  colonizing leftover space 101

immigrant children 47
love of nature 159
multiculturalism 140
Trafalgar Square 104
Wetland Centre 173
loose materials for play 65, 69–71, 73, 119, 121
Louv, R. 161, 163
Low, Tim 171

Maller, C. et al. 164
Malmö, Sweden 204–205
Manchester 107
Mand, K. 47
Māori people 149
maps and drawings 235–237
markets 124–125
Marseilles 140
Matthews, H. 20
Matthews, H. and Tucker, F. 233
Maxey, I. 118
Melbourne 31, 82, 101–102, 107–108
adult-dependent mobility 183–184
streetscapes 195–196
mental health 16, 80–81, 164–165
micro-spaces of the city 10
mobile phones 28
mobility 15, 181
see also independent mobility (CIM)
Moore, R. C. 10
Moore, R. and Wong, H. 66
multiculturalism 140–141, 152
Mumford, Lewis 95–96
Munich 247
museums and art galleries 128–132
Muslim children 147–148

national playground standards 60
National Society for the Prevention of Cruelty to Children (NSPCC) 49
natural world
community gardens 122–123
declining access to 10, 168–170
green cities 172–175
importance for children 161–163
necessary for health and well-being 164–165
nurturing a love of 159–160
opportunities to experience 165–166, 182–183
presence in cities 166, 170–172
re-engaging with the wild 175–176
threats to 160–161, 172
nature-deficit disorder 161, 251
Neale, B. and Flowerdew, J. 46
needs of children
contact with nature 163
with disabilities 118–120
housing 40–43, 248
mobility 15
protection and risk 60–61, 245–246
school grounds 65
universal 8
urban design 205–206
neighbourhoods 77–78
child-friendly 78–83, 87–88, 191, 193–196, 249
children's independent mobility 185
children's views on 89–90
community gardens 122–123
impacts of school closures 191
local markets 124–125
planning and design 83–87, 89, 210–212
segregation 142–143
service spaces 116, 133
neighbourliness 78–79
neoliberalism 20, 116–117, 190
Netherlands 210–213
see also Amsterdam
Newcastle-upon-Tyne 30
New Urbanism 210
New York 19, 140, 175, 247
New York City Streets Renaissance Campaign 34
New York (state) 122
New Zealand
child impact assessment tool 230
children home alone 49
children's views of their mosque 147–148
independent mobility (CIM) 186
Māori people 149
planning for children 6–7, 231
playground design 218
school segregation 143–144
see also Dunedin
New Zealand Social Report 44
Northern Ireland 150–151
Norway 62

obesity and diabetes 11, 26, 50, 165
O'Brien, M. et al. 187
Organisation for Economic Co-operation and Development (OECD), child poverty 43–44

organized sporting activities 24
outdoor play
  decline in opportunities 23–24, 48, 58
  experience of the natural world 159, 165–166, 168–170, 175–176
  fixed equipment 67
  greening of school grounds and forest schools 67–69
  loose materials 65, 69–71, 73, 119, 121
  organized 11, 24
  space at home 52
  and urban design 216–219
  see also sport

Pain, R. 25
Palmer, S. 175
paranoid parenting 7–8, 79
Paris 103
parks 5, 19, 117, 167, 217
  regeneration 173
  Vancouver 209
participation
  access modes 238
  benefits for children 233–235
  Bologna Children's Manifesto 205
  Canaansland 82
  design of school grounds 66
  difficulties around representation 233
  international support for 225–227
  methods 235–237
  public art 105–106
  research 238–239
  rights to 246
  working with professionals 224–225, 227–233, 239–240
passive surveillance 22–23, 27, 30, 191
peak oil 16, 193, 209, 243–244, 246
pedestrians
  prioritization 211, 213, 220
  street closures to traffic 97, 194, 247
permaculture playgrounds 122–123
Perth, Australia 173, 207–209
physical exercise and games 65–68, 154
  see also outdoor play
Piaget, Jean 3
place 4
  and culture 138
  sense of 183
  valued by children 20–21
planning and design
  building neighbourhoods 83–87
  children's involvement 31
  children's views on 89–90
  children as users 5–7
  globalization of building form 154
  for health and social well-being 80–82
  playgrounds 117–118
  principles for good neighbourhoods 87–89, 193–196
  school grounds projects 65–66
  service spaces 116–117
  to foster neighbourliness 79
  see also professionals; urban design
play
  changes in types of 24
  commercialized experiences 117
  emerging concerns 10
  housing needs to accommodate 42
  mixed age groups 70–71
  moving inside of 217–218
  and public art 104–105
  value of 62–63, 251
  see also outdoor play; playgrounds
Play England 61, 100
playgrounds 117–118, 166, 217
  adventure play 120–121
  for children of all abilities 118–120
  commercial 125–126
  design 218–219, 237
  natural spaces 175–176
  safety surfacing 26
  see also schools and school grounds
pollution 183–184
Portland, Oregon 42
Postman, N. 7
Prague 103
prisons 130
private sector
  play centres 125–126
  shopping areas 123–124
professionals 223–224, 229
  working with children 224–225, 227–231, 239–240
Pruitt-Igoe Estate, St Louis 83
public art 101–103
  children taking a role 105–106
  as interaction and adventure 104–105
  sculpture 98–99
public housing developments 83–84

public spaces
  city centres 97–100
  design for enjoyment 5
  see also service space
public transport 6, 189–190, 193, 246–247
  London 109
  Washington DC Metro 34

Quortrup, J. 13

race
  fear of victimization 25
  see also cultural differences; culture
Radburn layout 83
religion 146–148
  celebrations and festivals 79, 151–153
  and conflict 150–151
representation and participation 233
Rescue Mission Planet Earth 225–226
research
  action research 65
  participation in 238–239
resilience 16, 24, 29, 197
  in challenging neighbourhoods 82
  cities 209, 244–250
rights of children 119, 225–227, 246
  cultural 139–140
  in the home 52
  to public domain 10–11
Rio Conference (UNCED) 225–226
risk anxiety 60–61, 69
risk-averse behaviour 8, 24–27, 169
risk compensation 61
risks
  from limiting freedom to play 60
  natural world 163
  segregation and conflict 150–151
risk-taking in children 60–62, 163, 169, 252
  forest schools 68
Royal Town Planning Institute (RTPI) 228
rules 28, 51–52, 169, 192–193
rural areas, independent mobility 30–31

Sandercock, L. 140, 155
scale and distance 210
Schmelzkopf, K. 122
schools and school grounds 57, 59, 249
  alternative models 73
  children's views on 71–73
  choice of schools 63–64
  distance from home 190–191, 194, 227
  environmental learning 64–66, 73, 162–163, 172
  fixed play equipment 67
  greening and forest schools 67–69, 173
  loose materials and risk reframing 69–71
  physical exercise and sport 66–67
  reflecting change elsewhere 58
  religion 146–147
  segregation and inclusion 143–146
  value of play 62–63
sculpture 98–99
Seattle 59
sedentary lifestyles 11–12, 50
segregation and inclusion 155
  age 23, 117, 125–126
  conflict 150–151
  disability 145–146
  indigenous children 149–150
  race and ethnicity 142–143
  religion 146–148
  schools 143–145
semi-private space 42
service space 115–117, 132–133
  community gardens 122–123
  hospitals and healthcare services 94–95, 126–128
  local markets 124–125
  museums, art galleries and libraries 128–132
  playgrounds 117–121
  prisons 130
  shopping centres 123–124
shared care 46–47, 52
Sheffield 124, 189
shopping centres
  local 124
  malls 96–97, 123–124
SickKids 86–88
site selection and design 216–217
skate parks 12, 15, 116, 216
  child participation 231–233
Skelton, J. A. et al 165
Skenazy, Lenore 5–6, 26
slow cities 34
slum-clearance 84–85
Smith, F. and Barker, J. 31
Smith, Hingangaroa 149
snakes 163

social capital 78, 116, 185
social change and home relationships 45
social development of children 3, 70–71
  importance of neighbourhoods 87, 89
  key themes 8–9
  learning consequences of actions 61
  redesign of school grounds 66
  see also participation
social hierarchy among children 65, 70
social interactions and technology use 50–51
social isolation 78–79, 108
socially determined space 5–7, 14–15
social paradoxes 13
social polarization 16, 181
social traps 191–192, 196
socio-ecological model of healthcare 127–128
socio-economic status 29–30
SoftGIS 237
Soori, H. and Bhopal, R. 30
space
  children's use of 9–10
  importance of 4–5
  socially determined space 5–7, 14–15
  space-based paradoxes 13
Spirn, A. W. 172
Spock, B. M. 3
sport 66–67, 154
Stearns, P. N. 7
Steen, S. 73
stranger danger 22, 26, 28, 183, 191
  less fear in Germany 30
streets
  design and culture 5, 191, 194–196
  entertainment 97–98
  games 24, 78
  pedestrianized 97, 194, 247
  reclaiming for children 34, 182, 190
  retreat from 11–12, 22
suburban low-density housing 84–86, 207–208
surplus energy theory 63, 70
surplus safety 61, 70
surveillance 27–29
  shopping malls 123
  see also passive surveillance
sustainable development 160–161
  housing 214–216
  urban 204–205, 209
Sweden 83–84, 186
Sydney
  adult-dependent mobility 185
  community gardens 122
  Darling Harbour 5, 104
  loose materials in school grounds 69–70

technological development 152
technology
  space for 49–51
  to facilitate participation 236–237
terrace/row housing 85
third spaces 116
Thompson, Sarah 169
Thomson, S. 61
time pressure 21, 184, 191, 196
Timperio, A. et al. 31
Titman, Wendy 64–65
Tokyo 108, 247
tolerance of children 45, 100, 152, 169, 192
Toronto 140, 147, 155, 187, 252
tourism in cities 95
traffic
  generated by adult-dependent mobility 184–185
  generated by school closures 190–191
  safety 11, 25–26, 34, 183
  speed restrictions 190, 194
  and street relationships 212
transport modes 189–190, 193, 210, 246–247
Tranter, P. J. 185
trees 32–33
trust 28–29, 78
TV 48, 50–51

unaccompanied children *see* independent mobility (CIM)
United Kingdom (UK)
  changing relationships with nature 168
  child poverty 44
  children home alone 49
  Code for Sustainable Homes 215
  faith-based schools 146–147
  Home Zones 194–195, 212–213
  independent mobility (CIM) 185, 187–189
  inner city populations 107–110
  markets 124–125
  multiculturalism 140
  risk-anxiety 61

segregation 142, 150–151
see also individual cities
United Nations (UN)
Children's Fund (UNICEF), Child Friendly Cities project 12
Conference on Environment and Development (UNCED) 225–226
Convention on the Rights of the Child 119, 139–141, 225, 228, 240, 246
Convention on the Rights of Persons with Disabilities 119
Educational, Scientific and Cultural Organization (UNESCO) 145
Growing Up in Cities Programme (UNESCO-MOST) 9, 80, 82, 87, 89
International Year of Biodiversity 161
World Summit for Children 226
United States (US)
Boundless Playgrounds 120
Bus Initiative 145
child care arrangements 48
child-friendly hospitals 127
child poverty 43–44
children home alone 49
community gardens 122
faith-based schools 146
inner city populations 107
journeys to school 190
No Child Left Behind Act 63
play opportunities in schools 58–59
public housing developments 83
public transport 189–190
TVs in children's bedrooms 50
see also individual cities
universal needs 8
urban design 203–204, 219–221
child spaces 216–219
city level 209
design principles 205–207
examples of success 204–205
housing 52–53, 84–87, 191, 214–216
neighbourhoods 210–212
Perth, Australia 207–209
streets 212–213
urban form and transport 189–190
urban growth 203
urban wildlife 166, 170–172

Vancouver 108
van Zutphen, M. et al. 50
victimization 25
views of the world 31–32

walking school bus (WSB) 29, 34, 186, 196
Walsh, P. 217–218
Ward, Colin 6–7, 10, 115, 118, 217, 220
Washington DC 34, 192
Watson, S. 124–125
well-being 15–16, 164–165
Weller, Richard 207
Wellington 98–99
Wells, N. and Lekies, K. 162
Wheway, R. and Milward, A. 15
Whitzman, C. and Mizrachi, D. 31
Whitzman, C. and Pike, L. 192–193
wildlife experiences 159, 162–163, 166–167, 216
Williams, S. and Williams, L. 28
Wilson, E. O. 161
Wood, D. and Beck, R. J. 51–52
Woolley, H. et al. 100–101
working hours 45, 48, 208
World Conservation Union 161
World Wide Fund for Nature 205
A World Fit for Children 226
Wrexham, Wales 121

Yabe, N. 108
Young, M. and Wilmot, P. 84
youth councils 233, 238, 240
Ypenburg, Netherlands 211

zero tolerance 192

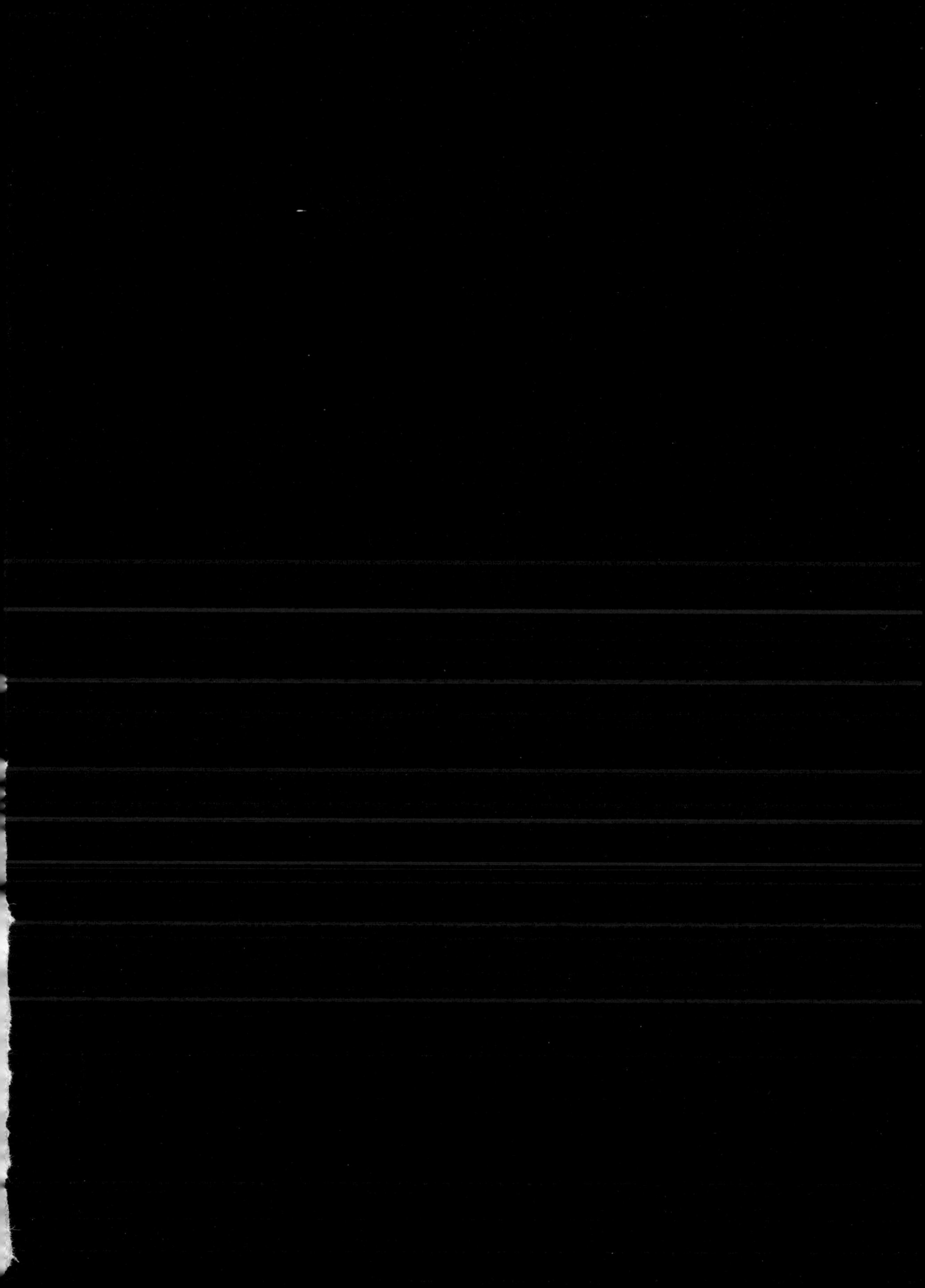